Science of the Soul
Soulutions of the Heart

A different lens through which we can view ourselves with greater clarity...

Science of the Soul
Soulutions of the Heart

*A different lens through which we can view
ourselves with greater clarity...*

Mary Kay Stenger, Ph.D.

Dandelion Books, LLC
www.dandelion-books.com

A Dandelion Books Publication
Mesa, Arizona

Science of the Soul: Soulutions of the Heart,
by Mary Kay Stenger, RN, CHt., Ph.D.

Hard Copy Version
ISBN 978-0-9967089-4-4
Library of Congress Catalog Card Number 2016946763

Front cover design by Accurance.com
Book cover layout and book interior design by Accurance, www.accurance.com

Dandelion Books & Dr. Mary Kay Stenger are extremely grateful to pexels.com for the free use of the Pleiades cluster photograph for the book cover ["All photos on Pexels are licensed under the Creative Commons Zero (CC0) license. This means the pictures are completely free to be used for any legal purpose. The CC0 license was released by the non-profit organization Creative Commons (CC)."—https://pexels.com/photo-license

Dandelion Books, LLC
Printed in the United States of America
www.dandelion-books.com

Mary Kay Stenger captures the heart of existence itself. In a world that seems constantly out of control and chaotic, she helps the reader to understand the importance of creating a healthy soul and peaceful solution-based foundation upon which we can depend. She explains beautifully how the healing of our spirit or soul can open the door to solving many of our daily trials.

This book has the potential to heal and enhance your life. In our personal search for happiness, understanding the importance of why we think, feel and act the way we do helps us to overcome life challenges. Mary Kay gives vision to why and how we can attain peace and balance to our lives.

Incredibly insightful, Mary Kay is nothing short of phenomenal in her smart, insistently scientific approach to giving the reader encouragement and guideposts to follow, that allow the reader to experience their "very own soul journey" and life adventure with passion and perspective.

Mary Kay has done immense justice to the earlier works of Gary Schwartz, Candace Pert, Elisabeth Kubler-Ross, Norman Cousins, Carolyn Myss and others, as she guides the reader in exploring the "mind-body connection" toward effective personal healing. To quote Mary Kay: "Once we understand the mind-body connection and the fact that we operate not from the conscious rational part of the mind but from that other hidden part... we will then learn how to tap into our inner power so we can change any aspect of our life that is keeping us from experiencing our greatest freedom, joy and abundance."

> —Dr. Paul McCarty, CEO, Nova Institute of Technology; psychologist & education specialist; college professor; researcher

---

Science-mindedness is an attitude and process of the mind, and Mary Kay Stenger illustrates this in spades. Mary Kay reveals how her creative science-mindedness takes her to the existence of the soul and shows us how to apply this knowledge in a practical way for our physical, emotional, and spiritual wellness. Well written, informative, inspiring, and useful—Science of the Soul is science for the soul.

> —Gary E. Schwartz, PhD., Professor of Psychology, Medicine, Neurology, Psychiatry & Surgery, and Director of the Laboratory for Advances in Consciousness & Health, University of Arizona; author of *The Afterlife Experiments, The G.O.D. Experiments, The Energy Healing Experiments,* and *Synchronicity and the One Mind*

Here is a woman who set out to engage in multiple realities to validate them for herself so she could become the compassionate caregiver that she sought in others. Mary Kay Stenger has paid a high price for the wisdom and healing gifts that emerged from her experiences.

This book is a treasure.

> —Robert J. O'Connor, CHt., Quantum Source Integration Therapist, www.wholeheartedenterprises.com

---

Dr. Mary Kay Stenger's latest book asks and answers questions many of us ponder every day: What do you really want? Is it in your soul's best interest? Will it bring you good health, contentment and joy? The author's "soulutions" provide a different lens through which we can view ourselves with greater clarity.

The book's premise is that all mind-body problems are soul issues and we are all eternal spirits in human bodies. Aristotle said it first—that the soul and body react sympathetically with each other. Through fascinating case histories of everyday people with common health disorders, Dr. Stenger makes her point in clear ways that are easy to understand. Questions, reflections, and activities at the end of each chapter are helpful and enlightening.

As a retired nurse, I particularly enjoyed the sections on near-death experience and the heart-soul connection because it emphasized what health care workers have known for decades. For example, the blood and circulatory system is the first organ system to form in embryos. No living creature can survive without the heart and circulation. None can thrive unless the circulatory system is healthy and strong. Stressful, negative emotions deplete humans and adversely affect all our body systems, especially the heart. The mental, emotional baggage we carry with us is exhausting so why do we insist on carrying it when our bodies contain a blueprint for healing? That blueprint is the soul. If we can believe that each of us is a unique eternal soul connected to the Divine and that everything we think or do impacts that relationship, the healing possibilities are endless. The near-death histories in this book prove the soul connection in hopeful, exciting ways.

Whether you are religious or atheist, spiritual or non-believer, cynic or optimist, you will find intriguing information in this book. If we can believe that everything we perceive is connected, we can improve lives, health, and the world. Dr. Stenger's message is laced with hope, truth, and genuine love and highly recommended for readers of all ages.

—Laurel Johnson for *Midwest Book Review*

What is the purpose of our life except for soul growth?

—Mary Kay Stenger

Dedication

To Michael G. Glynn, my soulmate, the man who opened my heart and touched my soul. You flowed into my life like a calm, refreshing stream and brought peace and love to our family.

May your soul rest in peace until we are fully reunited in God's presence again.

Acknowledgments

I wish to express my deep appreciation and gratitude to my family and clients. By sharing with me your life experiences, you have helped me evolve on my own soul path.

A special thanks to Carol Adler, my editor and publisher, for guiding and encouraging me every step along the way in writing the words of my heart.

Contents

Preface

If ever there was a time to address the question of a possible Sacred Partnership between us and them [departed souls], creatively and responsibly, this is it.

—Gary E. Schwartz, Ph.D., *The Sacred Promise: How Science is Discovering Spirit's Collaboration with Us in Our Daily Lives*

In 2010 after the death of my 29-year-old daughter, Sara, a friend introduced me to the work of Dr. Gary E. Schwartz regarding scientific validation of the existence of the soul or Spirit. Dr. Schwartz's book, *The Afterlife Experiments: Breakthrough Scientific Evidence of Life After Death,*[1] confirmed my own encounters with Spirit. Through those experiences I had developed a personal belief that:

1. Our essence or soul energy continues on or lives beyond our physical body.
2. Deceased individuals, e.g., family and other loved ones who are no longer in physical body often assist us during moments of our own physical life.

As I read Dr. Schwartz's book and learned more about his research, my heart leaped for joy. This was the solid evidence I had been searching for that would scientifically validate the basic premise of my healing practice, viz., that **all mind-body issues are soul issues.**

In 2011, another groundbreaking book by Dr. Schwartz, *The Sacred Promise: How Science is Discovering Spirit's Collaboration*

---

[1] Schwartz, Gary E., M.D., *The Afterlife Experiments: Breakthrough Scientific Evidence of Life After Death*. Atria Books, 2003.

with Us in Our Daily Lives,[2] provided further validation of the existence of Spirit and the soul.

Regardless of the nature of the issue that a client may bring to me, I approach and treat it at the soul or "source" level. I know that once my client begins to communicate and connect with their soul, they will have all the answers they need in order to start healing themselves and creating a joy-filled life that resonates with their deepest desires.

With science backing my personal soul experiences as well as those in the healing room, I was now confident I could pass on Dr. Schwartz's research about the soul without feeling like I was "preaching" to anyone. It was time for me to write a book that describes the healing methods I use with my clients and also explains why they achieve such powerful permanent results or what I like to call "soulutions."

The Science of the Soul opens windows of insight by giving us a different lens through which we can view familiar aspects of ourselves with greater clarity.

In this book I present some of the scientific principles of our Soul's eternal life. I show how we can use sound tools of healing that are based on these principles in order to live a more joyful, loving and fulfilling life.

Whether we are business executives, service providers, stay at home mothers, government workers, artists, athletes, students—whatever our profession and wherever we are on our life path—I believe we can benefit from an expanded perspective of who we really are by exploring the premise that the soul is our Personal Guidance System, or PGS. Our soul has all the answers and "soulutions" that will move us past any challenge, large or

---

[2] Schwartz, Gary E., M.D., *The Sacred Promise: How Science is Discovering Spirit's Collaboration with Us in Our Daily Lives.* Atria Books, 2011.

small, and any form of suffering that is currently consuming our will to have an exceptional life.

Dr. Gary Schwartz's work proves to us scientifically that none of us is alone, ever. No one has to live their life or pass to their death alone. We can use Dr. Schwartz's sound science for the growth, protection, and assistance of our own souls. The Science of the Soul or Spirit expands awareness of our individual Soul Path in this life and beyond.

It is my hope that you can use the information regarding the new science of heart, soul and mind-body medicine combined with the healing tools I present in this book, to create a solid soul-and soulution-based foundation for your daily life.

Introduction

In this book I present basic information about heart and soul science. I support these scientific findings with live examples and demonstrations that illustrate and describe a "soul path" or "soul journey." I also provide tools and assistance, as well as hope, for your very own soul journey. I wish to give each of you expanded perspective, insight and excitement about this adventure called life.

When someone asks me how I work with a client, I tell them I address their issue at the soul level. I help them find "soulutions" to their challenges by guiding them through a process that shows them how to communicate with their soul, the Divine essence of who they are.

Our souls already have the answers to whatever issues may be troubling us. Once the process of communicating with the soul is perfected, the client's conscious or "outer" self (ego) and their soul start to take the first steps forward on their soul path.

This process is really quite easy once we become open to it and allow our hearts to lead the way. Just as the soul is the cornerstone of our physical, mental, emotional and spiritual being, the heart is the center or fulcrum of our mind-body. **It is through our heart that we connect to our soul.**

Even the simplest problems have the potential for becoming "soulutions" for furthering the growth of our souls. Since the soul is immortal, a fact that has been scientifically proven, thanks to the work of Dr. Gary Schwartz,[3] we can take these soulutions with us after we leave our bodies, to benefit us "in the beyond."

---

[3] Ibid.

In addition to Dr. Schwartz's work, we have numerous clinical reports from those who have actually died, experienced life outside of their body and returned, having what we term a NDE (Near-Death Experience). These first-person reports are highly consistent and life changing.

Health professionals, psychiatrists, psychologists, and hypnotherapists can also cite thousands of cases in which they regressed individuals to past life experiences. Many of these individuals even have memories of life between life, often referred to as "LBL."

The common thread running through all of these experiences is greater peace, a broader perspective, genuine healing, and absence of fear of death.

Suffering Is Merely Separation

When we become disconnected from our soul, often we lose our way on the path of life. Fear, confusion, uncertainty and anxiety can lead to physical, mental, emotional and spiritual stress that manifests as illness, emotional pain or dysfunction. The problems persist and usually worsen until a person reaches a point of crisis.

A crisis is often an opportunity for soul growth. When a person cries out for help, the soul responds. At this point, either they start to listen to these inner messages and do whatever is necessary to reconnect with their soul, or they continue to suppress their pain and suffering. Usually the latter takes its toll in the form of early aging, illness, and dysfunctional behaviors, such as addictions, severe depression, physical death, and even suicide.

As a healer, I am merely the navigator or tour guide who assists my clients on their journey of reconnecting to their soul. Together we embark on a discovery process that ultimately identifies the source of their problem. Then we start to build the bridge between the "outer" and "inner" selves. This soul journey as I like to call it, profoundly affects every aspect of the person's life, giving them "soulutions" or soul-based solutions to the issues that have been troubling them. Usually a soulution-based life leads to personal transformation. This is authentic healing.

In this book, my goal is to use the scientific research of Drs. Deepak Chopra, Candace Pert, Carolyn Myss, Gary Schwartz, and others, to describe the dynamics of mind-body healing. I then use these findings to create a very personal mind-expanding experience for each of you.

In Chapter One, I explore what we mean by "science of the soul." I expand upon this in Chapter Two by describing the mind-body connection and scientific research that led to the development of a new field of mind-body medicine, technically referred to as psychoneuroimmunology.

Just as allopathic medicine is the scientific treatment recommended when addressing physical body issues, such as a broken bone or crisis care as in an automobile accident, hypnosis, hypnotherapy and regression therapy are the scientific treatments most often used when addressing mind-body or soul-connected issues. For those who are unfamiliar with these practices, in Chapter Three I provide a brief explanation of how they work and why they are such effective tools for re-connecting us with our all-knowing self. Throughout this book thereafter, I provide numerous demonstrations in which I used hypnosis and regression therapy in order to assist my clients in effectively achieving re-soulution of illnesses, dysfunctions and other forms of dis-ease. I also include exercises and visualizations that you may incorporate in your own daily mind-body healing routines.

In Chapter Four, I further explore the mind-body connection from a scientific perspective by presenting the work of Dr. Candace Pert, who is attributed with having discovered "emotion molecules," or neuropeptides. By understanding the dynamics of our body's cellular information system, it becomes apparent that each of us is in charge of our health and well-being as well as our attitudes and outlooks. Carolyn Myss's work as a medical intuitive supports Dr. Pert's findings through her discovery of our innate ability to tap into our mind-body's information system at the deepest level. It is this connection that opens communication between our conscious self, or our ego, and our highest or cosmic spiritual self. As I work intuitively with my client, I assist them in achieving clarity while processing new information that ultimately gives them the potential for effecting authentic healing and transformation.

Next, in Chapter Five, I present the science behind our daily actions as viewed from a mind-body perspective, and the dynamic behind soul healing. How does mind-body medicine perceive soul encounters? Can these non-local experiences be explained scientifically or are they only figments of the imagination? In Chapter Six, I provide additional scientific evidence for "soul encounters," and in Chapter Seven, I discuss Near-Death Experiences, citing considerable scientific proof that validates these visits to The Other Side.

In Part Two we are now ready to construct the bridge between Self and Soul, the mind-body and the superconscious, or the deepest eternal part of us. Modern science tells us that the heart, which is the body's primal organ, is our connection to the soul. In Chapter Eight, I discuss the heart connection from both an allopathic and psychoneuroimmunological perspective. The heart's electromagnetic field is the mind-body's most powerful rhythmic energy field. It is also an irrefutable scientific fact that the emotion of love, which vibrates at an extremely high frequency, is directly associated with the heart organ. Thus, it is no mystery that a person who is following their soul path and comes from a place of love will experience the heightened benefits of heart-and-soul-based living and healing.

In Chapter Nine, I discuss the soul connection in terms of religious beliefs and practices that either place obstacles on our soul path, or offer a clear and simple way to access the

Divine within. By allowing the Divine to guide our path and by maintaining a strong heart connection, we can be reassured of maintaining our balance and meeting our challenges with quiet inner strength.

Sometimes our life choices are not as easy or clear as we would like them to be. The road to bliss is frequently littered with pebbles, larger rocks and even detours that can distract us and take us off course. In spite of Divine guidance—mind-body-Spirit messages that are difficult to ignore—often we stubbornly choose the hard way or the long way around. Why is that? In Chapter Ten, I touch upon some of the reasons for this human conundrum. Actually, challenges are a major part of our soul experience. In Chapter Eleven, I discuss these challenges as gifts or opportunities from which we can grow. If we did not have problems to solve, we would have no way of strengthening our resolve and clarifying our intentions along the way.

Finally, in Part Three, "Creating Your Heart Path to Wholeness," we can start to apply what we have learned in Parts One and Two by opening to self-awareness and actually "becoming" our soulution. Chapter Twelve discusses ways to expand our awareness, first by accepting our authentic self and then by exploring our heart connection.

In Chapter Thirteen, as we continue to expand our awareness by opening our hearts, the magic and miracles that we experience are more than enough scientific evidence for the existence of the spiritual world. From that time forth, we start to see our life from a place of balance and inner peace. We literally "become" our soulution.

The Science of the Soul is an easy, loving way to launch your own exciting, creative life journey both in the "here and now" and forever after.

Part I – Soul Science

A New Way of Perceiving Our Divinity

Science has shown us that our subconscious mind sees and picks up what it is primed for. As you expand your awareness and **include awareness of soul as your center or the core of your being, I believe new opportunities will arise for you**. You may find more ease in day to day life and delight in the recognition that you are not navigating through life alone.

1

What is 'Science of the Soul'?

If the research of Dr. Gary Schwartz is true, his scientific discoveries can change everything in your life. If his work is not true and you consider it, it can still change your life.

The Miracle of Science

I love science. The first actual science experiments I conducted were in fifth grade when I was a member of the Camp Fire Girls. Our leader, Skip Flora, showed us how to look at a leaf that we plucked from a tree through a "lens" that revealed far more about the leaf than we could see merely with our natural eye. She instructed us to grind up the leaf and put it in a solution into which we placed a litmus paper strip.[4] The paper absorbed the liquid and a rainbow of color appeared before our eyes.

---

[4] Like most paper, litmus paper is made from wood cellulose. The wood is treated with solvents prior to paper manufacturing in order to remove resinous material and lignin from the wood. One of the most common solvents in the United States is a sulfate—either sodium sulfate or magnesium sulfate. The ability of litmus paper to change color when exposed to an acid or base is a result of litmus paper being infused with lichens. In the plant world, lichens are unique in that they are actually two distinct organisms, a fungus and an alga, living as one. Botanists classify lichens as fungi because it is the fungi that are considered to be responsible for sexual reproduction. http://www.madehow.com/Volume-6/Litmus-Paper.html

This experiment was magical to me. I became hooked on science and its ability to show us a world that may not be apparent to us until or unless we examine it in greater detail. I further realized that we needed to know what questions to ask, e.g., what elements were in the leaf that caused the litmus paper to change color? Why did we see a prism of color? Would these colors also appear in a leaf that was no longer green, that had fallen from the tree long ago?

I still love science. In college, I changed my major five times in five semesters, from one science to another. From a scientific perspective, I wanted to know everything about everything! I knew I needed to be practical, however, so in my sixth semester, I began a nursing program at the Los Angeles County Medical Center. As the oldest of seven children and self-supporting in college, it made sense to obtain the necessary credentials in a field where I knew jobs were always plentiful.

What I didn't know at the time and what I learned later was that by choosing to become a nurse and engage in the study of physical human science, I had already started to build the bridge between the scientific or diagnostic approach to disease according to Western medicine, and healing as it is intuitively known through spiritual awareness of our higher self. My questioning mind combined with daily life and death experience as a nurse motivated and inspired me to continue to study the emotional and spiritual components of the human mind-body.

Some of you may be puzzled by the title of this book. You may be asking, "Is there really such a thing as 'science of the soul'? Isn't science based on hypotheses that use microscopes, test tubes and Petri dishes to prove something is valid or true?

The discussion seems to center around the term "soul" or "Spirit," referring to that part of the world that is invisible and for which there is no proof of its existence. If one compares it to other invisible phenomena, such as "gravity" or "electricity," for example, there is heavily documented proof. The former falls into the category of religion or spirituality, and the latter belongs to the category of science. Oil and water, science and the soul do not mix. Therefore, it follows that a Science of the Soul is impossible.

Or is it?

If we are affiliated with a certain religion or spiritual group that teaches us about "God," then we have been taught from birth that we don't have to prove His/Her existence. Faith is all that is needed. Science belongs in the laboratory and soul talk belongs in a church, synagogue, mosque or meditation room.

Since this is what we have been taught to believe, it's no wonder you may be puzzled when you see the words "science" and "soul" or "Spirit" linked to each other.

'I Think; Therefore I AM'

In the 17th century, philosopher René Descartes established with absolute certainty the duality between science and spirituality, or religion.[5] Man is a rational being and God is a spiritual being. Since that time, we have maintained a separation between science or secular matters ("matters of the State") and religion (the Church) or spiritual matters.

Descartes's statement, "I think; therefore, I am," is called an *a priori* truth because it can be deduced by logic or reasoning. We would not be aware of our existence if we were not self-conscious, says Descartes. This is a logical deduction.

---

[5] Once the conclusion [that "I exist" and "God exists"] is reached, Descartes can proceed to rebuild his system of previously dubious beliefs on this absolutely certain foundation. These beliefs, which are re-established with absolute certainty, include the existence of a world of bodies external to the mind, the dualistic distinction of the immaterial mind from the body, and his mechanistic model of physics based on the clear and distinct ideas of geometry. http://www.iep.utm.edu/descarte/

We cannot apply the same logic to the existence of God, the Divine or Our Source, however. Up until this time, the only way we knew God existed was because our parents, a priest, minister, rabbi, shaman or other individuals whom we trusted, told us so. Belief in God was based on proclamations of blind faith. Belief in the fact that water turns to ice if its temperature is lowered below 32 degrees Fahrenheit was based on double-blind studies in the laboratory.

The Illusion of Duality

Long before Descartes and even before Plato and Aristotle brought forth their dualistic theories of realism and idealism, ancient societies were keenly aware of the mind-body connection. They also knew that all forms of life are integrally, organically and holistically connected to the cosmos through that part of us known as the soul. The soul was considered the cornerstone of all knowledge about ourselves and the world in which we live.

Modern science, academia and religion changed all of that and introduced a set of beliefs purporting a separation between "church and state" or secular and spiritual matters. Issues of God and spirituality were designated as the church's domain. Issues pertaining to the way the world works were assigned to philosophy, mathematics and the sciences. "Eternal God" lived in the heavens above and "mortal man" lived on earth below.

For many people today, this separation continues to hold true. If you attempt to discuss issues of the soul with a scientist, you may find this territory off limits and the discussion itself will be considered frivolous.

Nevertheless, any scientist will be quick to tell you that science can only explain *how* something works. It cannot explain *why* it works. As Thomas Edison demonstrated, if we turn on a switch connected to a light fixture, we will experience physical illumination. Is this not magical—and highly mysterious?!!

Likewise, if we connect our thoughts and feelings to the deepest part of ourselves known as the superconscious or higher mind, besides feeling a physical release, we will also experience mental, emotional and spiritual "illumination." Millions of people could not be imagining spiritual encounters, nor could they conjure up such a large number of extraordinary instances in which the soul plays an active role in their lives.

What if we were to bring Spirit/Soul into the laboratory and perform double-blind studies on the hypothesis that its existence can be proven after all? Could this belief in separation between church and state, science and spirituality, be the malaise that is at the root of all the chaos, uncertainty and confusion that is troubling today's world?

Writes Dr. Schwartz in *The Sacred Promise*:

> What if our increasing feelings of emptiness, loneliness, hopelessness, and meaninglessness are fostered by our belief in a Spirit-less Universe? ... What if Spirit is actually all around us, ready to fill us with energy, hope, and direction if we are willing to cooperate with it?"[6]

> ...If we, in our essence, are spirits too, can we come to see this possibility when we look at ourselves in the mirror? And can we draw on this great inner potential and power with wisdom and love to change our ways accordingly, before it is too late for humankind?

> I believe that science not only can address such questions, but in the process can potentially help increase our ability to receive spiritual information accurately and we can then act upon it safely and wisely.[7]

---

[6] http://www.amazon.com/The-Sacred-Promise-Discovering-Collaboration/dp/1582702586

[7] http://www.drgaryschwartz.com/

States Kristine Morris in reviewing *The Sacred Promise*:

> Gary Schwartz, Ph.D., brings his impeccable credentials as a scientist to these questions in what he calls "the science of the seemingly impossible"—the search for scientifically verifiable evidence of Spirit's interaction with us.
>
> In *The Sacred Promise*, Schwartz takes us into the science laboratory and its systematic, controlled experiments, as well as into what he calls "self-science," the data collected in the natural laboratory of daily life, to observe his explorations of the connection between the human world and that of Spirit.
>
> In our time, the fact that something is invisible to us no longer means that we disbelieve that it exists—all of our newest technologies are based on invisible electromagnetic frequencies, and new measurement devices are revealing that the universe is filled with light frequencies to which we are, literally, blind. In spite of this, it is still considered taboo to scientifically research the spiritual realm. *The Sacred Promise* takes us beyond where most dare to go—to evidence for the concept that Spirit has both the active desire and the ability to interact with us and to help us. Considering himself an "orthodox agnostic" when it comes to afterlife research, Schwartz says, "For me, scientific integrity means following the data where it takes you. If there is a greater spiritual reality, and if it can actually assist us in our individual and collective lives, then the sooner we can receive this assistance the better it will be for our species and the planet as a whole. If ever was a time for us to awaken to these profound possibilities, this is it."[8]

Either we can believe or not believe that the soul exists. When it comes to the fact that electricity exists, we have no choice but to accept its existence because we can witness the effects. Yet is this not also true for the existence of the soul or Spirit? Have we not also witnessed the effects of Spirit?

---

[8] *Spirituality&Health,* May-June, 2011, http://spiritualityhealth.com/reviews/sacred-promise

What if we were to admit that we can describe electricity and how it works, but we don't know why—and likewise, that we can describe the soul and how it works but we don't know why? Could we not conduct the same rigorous scientific experiments on Spirit and spirituality as Thomas Edison and other scientists and engineers conducted on electricity?

What if I were to tell you that we have proof of countless well-documented cases describing actual communication with the soul, validating its presence and even its intervention? Could we not say then, that both electricity and Spirit fall into the same category *that one could call the mystery and miracle of the Cosmos?* Is this not the great mystery of the mind, whose intricacies can be placed under the microscope and described and recorded in terms of data, using the same scientific terms as for gravity and electricity?

The Reality of Illusion

Albert Einstein said, "Reality is merely an illusion, albeit a very persistent one."[9] This is a sagacious way of saying that quantum physics that taught us that the "reality" to which Descartes refers is not as clear and direct as he would have wished. The fact is, our rational mind and common sense are simply not capable of understanding the true nature of reality, as in the experience of a Beethoven symphony or any other great work of art.

Therefore, we can say that the greatest reality of all is the illusion of separation. This illusion is so *real*, we cannot even make the statement, "I think; therefore, I am," without acknowledging our existence, because the soul is our *core consciousness.* To quote Deepak Chopra, "The soul is the source of all our lives. It projects as the mind, the body and the universe of our experiences."[10] Although Descartes would like us to believe that we can carve the soul out of the equation altogether, it is just not possible.

---

[9] http://www.brainyquote.com/quotes/quotes/a/alberteins100298.html

[10] http://intentblog.com/deepak-chopra-what-is-the-soul/

Carolyn Myss, a well-respected medical intuitive, takes this even further and says, "I believe that each of us is guided by a Sacred Contract that our soul made before we were born. That Contract contains a wide range of agreements regarding all that we are intended to learn in this life."[11]

Without the soul, we cannot think or even be aware of our existence. The dichotomy is an illusion and the so-called "reality" of science is also an illusion *unless* we acknowledge that the unified field of consciousness or core consciousness is all there is. Whatever we deem to call reality in order to apply scientific principles to explain or describe it, cannot exist without our perception or awareness of it.

One could paraphrase Descartes and state: "I think *because* I am." The "I AM" is our soul, and the science of the soul is the application of scientific method to the phenomenon of soul—its characteristics, behaviors and outcomes, or consequences when tested and measured in various controlled environments.

Soul Healing or Science of the Soul Leads to Permanent Soulutions

The depth and breadth of Soul Science—healing at the soul level and approaching this process from a scientific perspective—can go beyond anything we have ever thus far seen or heard. It is limitless and ever expanding.

Once we understand the mind-body connection and the fact that we operate not from the conscious rational part of the mind but from that other "hidden" part, we will have acquired the two basic pieces of information that allow us to understand from a scientific perspective why we think, feel and act the way we do. We will then learn how to tap into our inner power so we can change any aspect of our life that is keeping us from experiencing our greatest freedom, joy and abundance.

_____

[11] http://www.myss.com/free-resources/sacred-contracts-and-your-archetypes/

The doorway to this process of self-discovery is the heart, which ultimately connects us to the soul. Through authentic soul communication, we discover what we truly desire and we start to resonate with who we really are. At this stage of soul healing or "becoming wholly connected to our Divine essence," transformation occurs.

Carolyn Myss said: "You can pretend to be something other than who you are but eventually you will run out of energy to continue because that's not authentically you."[12] Self-pretense, or trying to be the person you think you should be, also blocks intuitive ability. This prevents you from experiencing peace and inner joy at the deepest level.

I believe soul science is the most significant life-enhancing and eternally important science we can pursue. Moreover, it is a study anyone can undertake. Since we are all souls having a human experience, why would we not want to understand this deeper, wiser part of ourselves? In the bigger picture of life, it is this soul part of us that matters most.

As I mentioned earlier, the science of Spirit or Soul is actually ancient. The only difference now is that we are actually using Scientific Method to document what humans have experienced since the beginning of recorded time.

The Science of the Soul can:

1. Help you tap into your deepest consciousness and wisdom.
2. Assist in placing day to day life in a larger, eternal perspective.
3. Open your mind to the help that is there for you beyond this physical world.
4. Give more meaning to your every breath of life.

---

[12] Myss, Carolyn. *Archetypes: Who Are You?* Carlsbad CA: Hay House, 2013.

5. View day to day problems as opportunities to deepen your awareness and understanding.

A laboratory scientist collects data and draws conclusions about the probability that a given phenomenon or explanation might be true. In statistics, this is called "rejecting the null hypothesis."

In Soul Science, examples of given phenomena are: 1) actual communication with individuals who have left their physical body, 2) voices, signs, synchronicities and other clues that these deceased individuals are present with us, even though they are not in their physical body, 3) dreams in which deceased or physically invisible individuals appear or deliver messages to us, 4) voices or internal communication that move through us and deliver information or messages that we have not consciously created ourselves, 5) actual appearance of deceased individuals, sometimes fleeting or seen as shadows, and 6) uncanny miraculous protection.

Based on copious amounts of data in all six of these categories that I had already collected in my personal life as well as in the healing room, Science of the Soul was surely an accurate description of my work and also validated my life experience.

I can recall many occasions when I realized my life was protected or spared. I knew there was something beyond what my natural eyes could see that was working with me and protecting me. I'm sure you can also recall similar experiences. Through the years of my work and training, I have learned to use what I call my "inner eyes" or inner vision. With my inner eyes, at times I can see in my mind beyond this earthly plane.

Years ago, I was driving home from San Diego where I had taken my boys and some neighbors on a surfing trip. The Volvo station wagon was fully loaded with six 13-year-old boys and surfboards.

Recently I had purchased new tires; I bought the best ones that money could buy and didn't mind paying more for them. My car was usually loaded with children and safety was my top priority. The salesman told me that the tires I bought were built never to have a blowout. Instead, if the car hit an obstacle, the tire would slowly deflate. He said they were the safest tires on the market.

On this particular afternoon heading north to San Luis Obispo, I was feeling unsettled and restless about the long trip ahead, with so many teenagers and surfboards in the car. I had made this drive numerous times before and had never felt this way. I even considered waiting until the next morning to leave, rather than driving through the night—but that wasn't an option.

After driving 30 miles, I was still feeling unsettled, so at the San Clemente exit I pulled off the freeway and told the boys we needed to have a prayer for safety.

For some reason I kept thinking that I needed four strong angels, one to stand guard by each tire. In my mind's eye, I could see those mighty angels hovering around the tires; it seemed like they were slightly lifting the car. Logically, I knew I had new tires and a safe car, yet I was acutely aware that I needed additional backup for safety. As I drove through the evening into the night, I continued to ask for that protection.

I took a shortcut through a mountain pass because it was getting late, I was tired and I wanted to be home by midnight. We had moved along at a pretty good clip on the freeway, and now in the mountains, I kept seeing the four angels traveling with me as the boys slept. I wasn't sure why they were there.

As soon as I arrived in my hometown I felt a sense of relief, and as I slowed down to make the final turn onto my street—the only time during the entire 300-mile journey after leaving San Clemente that we slowed down—I heard a loud *BANG*. It sounded like an explosion!!

A front tire literally blew up.

The boys awoke with a start. We weren't far from our house so they were able to walk the rest of the way, carrying their surfboards.

I waited until the next morning to call a road service to change the tire. When the driver looked at the scattered remnants of the tire, he shook his head in disbelief. The tire had completely blown apart and pieces of it were scattered all over the road.

The next day when I walked into the tire store with the fragments of one of my very expensive tires, the tire salesman didn't even bat an eye. At once he gave me four brand new tires.

The following day I found an article in the newspaper about a recall from the tire manufacturer for the same tires I had bought.

Had that front tire exploded while driving 70 mph on the freeway or on the mountain pass, I don't know what would have happened. I do know that those four angels got us through 300 miles of night driving to a place of safety, where they finally let me slow down to a crawl before that tire blew.

I then understood why I kept seeing those four mighty angels literally carrying us home.

This was not the first time I had seen angels. Maybe they would have been there even if I hadn't asked, yet because I had been aware and had asked, seen and felt them, I had a deeper appreciation for our safety. I knew beyond any doubt that the safe timing of the blowout was within only a few moments of having happened while we were still on the freeway.

Questions & Reflections

1. How would you explain soul science to someone who has never heard the term before?

2. How would you demonstrate the existence of the soul to a person who does not believe in the existence of anything without physical form?

3. What is meant by "believing is seeing?" How does this pertain to the science of the soul?

4. If you follow the scientific premise that you are a soul living in a human body, what does that concept mean in your day to day life?

5. Can you recall a time when you recognized you were protected? Did this circumstance defy logic? If so, explain.

6. What was your last "lucky break" or "synchronicity"? Did you recognize it for what it was? Do you think maybe it was more than just "luck"?

7. Think about a time in your life when something occurred that defied reason; for example, a near accident, or possibly meeting someone who changed your life.

8. Have you ever experienced any "epiphanies" or moments you cannot explain that were so profound and left you with such a lasting impression, they have been a part of your consciousness ever since? Explain how they may have affected or influenced your life.

Exercise

I have included exercises throughout the book for you to practice. We will always begin by quieting the mind and body through attention to our breath.

Focus on your breathing for several minutes, drawing your attention inward with your breath and creating calm and peace in your body. Simply breathe in a sense of calm and exhale all other thoughts. Be present in this very moment.

Begin to focus on a specific situation that may have defied logic. Allow your view of the situation to run through your mind. Watch it as if you are watching a play or movie.

Now imagine you have a set of eyes that can see beyond this earth plane. Watch the same scenario and see if you can view any other additional information that helps you to fill in the missing pieces of the story. Trust the information you receive.

As you practice quiet meditation and seek information, you will find that it comes to you more easily and regularly.

2

The Mind-Body Connection:
Psychoneuroimmunology

From a holistic practitioner's perspective a person cannot be "broken."

Over the centuries, much attention has been paid to the influence of mental and emotional processes on health and disease. Aristotle was the first to suggest the connection between mood and health. He is credited with saying:

> Soul and body, I suggest, react sympathetically with each other. A change in the state of the soul produces a change in the shape of the body and conversely, a change in the shape of the body produces a change in the state of the soul.[13]

Yet only in the 20th century have scientists had tools powerful enough to measure the links between mental and emotional processes and to demonstrate that the immune system could be trained to use these links to change behavior patterns.[14]

In the 1920s, Russian scientists showed that classical Pavlovian conditioning could both suppress and enhance the

---

[13] http://www.azquotes.com/quote/1222188

[14] http://www.apa.org/monitor/dec01/anewtake.aspx

17

immune system. Psychologist Robert Adler coined the term "psychoneuroimmunology" and in the 1970s performed a series of ground-breaking experiments with rats that demonstrated the ability of emotional cues to alter the rats' physiology. He gave the rats an immune suppressing drug flavored with sweet-tasting saccharin. Eventually, they became so conditioned to the effects of this drug the saccharin taste alone caused a suppression of the immune system.[15]

Although these studies showed that the immune system could be conditioned at the subconscious or autonomic levels, it was not until the late 1900s that scientist Howard Hall demonstrated that the immune system could also be consciously controlled.[16] At Case Western Reserve University in Ohio, Dr. Hall experimented with self-regulatory practices such as self-hypnosis, biofeedback, guided imagery, relaxation and autogenic training. Using several control groups, Hall demonstrated that with conscious preparation, through using one of the types of practices noted above, individuals could consciously control the stickiness of their white blood cells as measured by both blood and saliva tests.[17]

Studies using the Wim Hoff Method for consciously controlling immune responses also produced interesting results. Wim Hof, a Dutch world record holder who is famous worldwide for his ability to resist cold, was commonly nicknamed "The Iceman":

> Scientists led by Matthijs Kox of the Radboud University Medical Center in Nijmegen, the Netherlands, studied his [Hof's] method, which is somehow similar to the Tibetan Tummo technique (yoga of inner heat) and involves third-eye meditation, breathing exercises and cold exposure, and used it to train 12 volunteers to fend off inflammation.

---

[15] http://www.healingcancer.info/ebook/candace-pert
[16] http://online.liebertpub.com/doi/pdf/10.1089/107628004773933334
[17] Op. Cit., http://www.healingcancer.info/ebook/candace-pert

In the framework of the study, 24 volunteers, including 12 people trained in the Wim Hof method and 12 who did not undergo any training, were subjected to the inflammation test, during which they were injected with a strain of bacteria that provokes flu-like symptoms.

As a result, volunteers who underwent training with the Wim Hof method reported fewer and less intense flu-like symptoms than those who did not. At the same time, trained volunteers produced smaller amounts of proteins related to inflammation, and higher levels of interleukin-10, an inflammation-fighting protein.

"Hitherto, both the autonomic nervous system and innate immune system were regarded as systems that cannot be voluntarily influenced. The present study demonstrates that through practicing techniques learned in a short-term training program, the sympathetic nervous system and immune system can indeed be voluntarily influenced," writes the scientific journal PNAS, where the study was published.[18]

When thoughts are negative, fear-based, or stress-induced, the body responds by turning off the immune system, but if the conscious and subconscious mind powers are aligned with positive thoughts, the body responds with enhanced healing and health. [19]

If the immune system can be altered by conscious intervention, what does this mean for treatment of major diseases such as cancer?

We know the central nervous system, like the immune system, has memory and the capacity to learn. The brain is extremely well integrated with the rest of your body at a molecular level. The concept of a network, stressing the inner-connectedness

---

[18] http://themindunleashed.org/2014/07/science-behind-consciously-controlling-your-immune-system-mind-body-connection.html

[19] http://undergroundhealthreporter.com/subconscious-mind-power/#ixzz405CFdRTW

of all systems of the organism, seemed not only possible but probable.

Scientific evidence of the systemic connection between the mind and body has the potential to shift the entire paradigm of medicine from a single dimensional Newtonian model to one that is both multi-dimensional and "quantum." The patient does, in fact, participate in the process of healing and recovery.

The Hypothalamic-Pituitary-Adrenal Axis

The hypothalamic-pituitary-adrenal (HPA) axis plays an important role in stress-induced immune-brain interactions. The hypothalamus, pituitary, and adrenal glands secrete hormones that assist our biological processes: digestion, immune function, sexuality, mood, etc. *Medical News Today* explains:

> One chemical of note involved in the HPA axis' work is corticotropin-releasing hormone (CRH). The hypothalamus releases CRH in response to stress, illness, exercise, cortisol in the blood and sleep/wake cycles. It peaks soon after waking and slowly declines throughout the rest of the day. In a stressed individual, however, cortisol levels are elevated for prolonged periods of time.
>
> During stress, the body believes it is in imminent danger, so cortisol triggers a number of metabolic changes to ensure that enough energy is available in case a fight or flight is necessary. One of these energy-saving tactics is to suppress the metabolically expensive immune system, saving vital glucose for the approaching life-threatening event...
>
> In this way, ongoing stress can reduce the capabilities of the immune system as the body saves its energy for a physical exertion that never comes.[20]

---

[20] http://www.medicalnewstoday.com/articles/305921.php?page=2

"I Want to Go on a Cruise!"

"Suzette," a middle-aged woman who worked as a nurse, was undergoing cancer treatment. She came to me with two goals in mind. "First," she said, "I want to be able to work throughout my chemotherapy. Can we minimize the side effects of the chemotherapy? Second, my white blood count is dropping very low. If I can raise my white blood cell count during my treatment to 3.5, I can cut down the days between treatments. I want to go on a cruise in January. I need to boost my immunity in spite of chemotherapy."

Suzette gave me the exact number, down to the specific tenth, where her white blood cell count needed to be, that would satisfy her requirement for going on the January cruise.

During the next two months, Suzette maintained her desired work schedule. A couple of months later she brought me the lab slip that was to determine whether she could cut back on her timing between chemotherapy sessions. Results on the slip would also determine whether or not she could go on the cruise.

Her lab slip—3.5—was exactly what she needed. It was down to the exact tenth of the number she set in her mind!

How did her body know what to do to make a clinical lab slip passable for her goal? Both Suzette and I were in awe of the wisdom and determination of her body.

In the mid-1970s, from the time I started my career as a nurse, I have watched this evolution of what is now known as mind-body medicine or psychoneuroimmunology. This has been both an exciting and sometimes frustrating experience, especially in the early days when I was surrounded by practitioners who had been trained in allopathic or what is known as Western medicine, with little or no experience about the mind-body connection when diagnosing and treating their patients.

After my first day on ward rounds at LAC-USC Medical Center, I vowed to myself that there had to be a better way to care for people by *preventing disease* rather than dealing with the suffering and ravaged bodies I saw on those medical wards. Even as a student I was already aware that our medical focus should be on health rather than disease.

From the perspective of disease, most of the cases I tended in that antiquated understaffed hospital could have been prevented. Clearly these patients were suffering from lack of proper nutrition, alcohol and drug abuse, and wounds from gang violence. (In those days, "gangs" were referred to as "Families feuding," often all the way up the ambulance ramp!)

I received excellent training at the Medical Center and was often told that some of the cases I saw were so rare, I may only encounter a few similar ones in my lifetime.

My first nursing job was at a Seventh Day Adventist Hospital, where I learned about the value of nutrition and had an opportunity to practice compassionate care of the dying. We had numerous lung cancer patients, the majority of whom had been heavy smokers. This was another example of unnecessary suffering, most of which could have been prevented if the patients had stopped smoking before these conditions became acute.

Eager to learn about disease prevention, I read more than 20 of Dr. Bernard Jensen's books on nutrition, iridology, and colon cleansing, and longed to study with him one day. Fifteen years later, I had this opportunity. Dr. Jensen had a healing clinic in San Diego and was experiencing great results detoxifying cancer patients so their bodies could heal. His philosophy and procedures made sense to me. Even in the early part of the 1980s, when most Americans had no knowledge of the mind-body connection, I believed in our bodies' innate healing wisdom.

In Washington, D.C., while working in the ICU, CCU, NICU and Emergency Rooms in the metropolitan area, I finished a Bachelor's degree in Health and Preventive Medicine at George Mason University. During this time, while studying how to keep people healthy, once again I witnessed large numbers of acutely ill patients.

After adopting my first child, I worked part time at St. Louis University Hospital, where they were performing some of the first heart transplants. These were my most critical patients. I knew that heart disease must involve a mind-body connection; I did not want anyone to have to go through this procedure if it could be prevented.

Two years after delivering quadruplets, while living near the Sacramento, California area, I returned to school and finished my Master's Degree in Health Education Psychology. I didn't want to leave my babies, but with one semester to complete and with the penalty of having to re-enroll in the Master's Degree program and start over, it seemed like the best decision to complete my degree at that time.

I had excellent professors and I knew then that mind-body medicine was exactly what I wanted to pursue when the children were older. At that time, there was still not much science supporting the mind-body connection, but the amazing results I witnessed were enough to inspire me to keep learning.

Next I trained extensively in hypnosis and hypnotherapy while also finishing my formal nutrition study with Dr. Jensen. When my youngest child was old enough to attend pre-school, I opened a small private practice specializing in hypnotherapy.

After my divorce, as a single mother of six children, I enrolled as a Ph.D. candidate in Psychology and finally completed my degree several years later when the older five children were no longer at home. Motivational Psychology using hypnosis became my specialty and I continued to increase my client base.

From a mind-body perspective, I began to see tremendous healing in my clients. I also noticed that as these individuals healed physical and emotional issues they seemed to move forward into deeper more soul-related ones. At the beginning of this mind-body-spirit work, I was as surprised as my clients at the rapid healing they were experiencing. Sometimes they reported that their doctors were baffled by their disease reversal.

As I became more energy sensitive, my intuition became an extremely active part of my work. I could feel the energy of my clients and see pictures in my mind to give me direction. I was exhilarated by the results, but at the end of the day, I was usually exhausted.

By this time, I had recognized that all healing was soul healing and I began to practice soul-centered healing using hypnosis and hypnotherapy. Healing was now occurring more easily and at an even deeper level. As the deeper soul-related issues healed, the body followed suit.

Two-and-a-half years of formal psychic and intuitive training and practice with experienced teachers gave me more confidence to believe in the information I received from my intuition. I was able to work even more efficiently and effectively while also maintaining my normal energy levels. I then understood that *everything in the universe consists of energy,* including our souls.

As I mentioned in the Preface, after the death of my oldest daughter at age 29, I had an opportunity to meet Dr. Gary Schwartz and witness his soul sensing equipment in action. Now I finally had solid science to back my emotional work as well as the soul work that is everyone's path in this lifetime.

Although I had seen angels and was aware of their assistance in the past, these occurrences started to become more frequent. I could now freely talk about the science of the soul and teach my clients to seek soulutions, instead of just solving problems. Problems became gifts to be used for growth of the soul and for shifting that person's perspective. No issue seemed insurmountable. In amazement I witnessed some of my clients turn tremendous pain into strength and healing.

Since we are souls living in a human body, the body is an important component of a soul-based life. Sometimes it is the body wisdom that can lead us to the soul.

A Brief History of Mind-Body Medicine

Mind-Body medicine is often called holistic healing because it addresses the total body, i.e., its physical, mental, emotional and spiritual aspects. Perhaps the greatest difference between allopathic and mind-body medicine is that allopathic medicine is disease-oriented and holistic medicine is wellness-oriented. A holistic practitioner believes that humans are "energy beings" that consist of various energy forms vibrating at different frequencies. An allopathic doctor views the human as a physical being with parts similar to those of a machine.

From a holistic practitioner's perspective, a person cannot be "broken." When they experience presenting symptoms, they are merely experiencing weaknesses or blockages in their energy systems. Once the blocks are removed or the weaknesses are strengthened, the issue will disappear.

In order to determine the nature of the weaknesses or blocks, the holistic practitioner considers the source of the issue, whether physical, mental, emotional or spiritual. An allopathic doctor, on the other hand, treats presenting symptoms that have manifested in the physical body in the form of pain, a sore throat, headache, skin growths, swelling, etc. As if dealing with a faulty car battery or radiator, using drug therapy or surgery, they recharge or replace the part that is defective.

In the past, when dealing with a person who had heart disease, doctors would possibly prescribe blood pressure medication, a drug for cholesterol, or other medications. Recently they have started to add preventive measures as well as suggested lifestyle changes to their list of remedial treatments, yet they are still bound by the dictates of their profession and training. Some physicians still limit their discussion of mental, emotional or spiritual (subtler energy) issues.

If the problem is mental, the patient is referred to a psychiatrist, who usually treats the issue with drugs. Most of us are aware by now of the side effects of antidepressants and other mind-altering drugs, so we needn't dwell further on this issue. In certain circumstances, drugs can be appropriate. In other situations, however, instead of solving the problem, they can complicate matters. Ultimately, the underlying problem remains unaddressed.

Fortunately, thanks to many dedicated medical doctors and scientists whose discoveries have delivered statistics that point out the interrelationship of the mind, body, and spirit, this paradigm is shifting. These findings confirm that mental, emotional and spiritual issues directly influence a person's physical health and well-being, particularly in the areas of:

- High blood pressure
- Asthma
- Coronary heart disease
- Obesity
- Insomnia
- Anxiety
- Diabetes
- Stomach and intestinal problems
- Menopausal symptoms
- Sleep issues

The following are some interesting statistics regarding mind-body medicine:

- ✓ According to the World Health Organization (WHO), if the major risk factors for chronic disease like poor diet, inactivity and tobacco use could be eliminated, at least

80% of heart disease, stroke, and type 2 Diabetes would be prevented as well.

✓ Women who participated in a mind-body program for stress reduction while undergoing IVF treatment have a significantly higher pregnancy rate than those who do not (52% versus 20%) (A. Domar et al., Fertility and Sterility, 2011).

✓ More than a third of Americans use some form of complementary and alternative medicine (CAM). Of this figure, 1 in 30 are referred by a medical service provider (A. Nerurkar et al., Archives of Internal Medicine, 2011).

✓ Bacteria residing in the gut influences brain chemistry and behavior. These findings are important because several common types of gastrointestinal disease, including irritable bowel syndrome, are frequently associated with anxiety or depression (P. Bercik et al., Gastroenterology, 2011).

✓ A 2011 study linked Metabolic Syndrome to memory loss in older people. Metabolic syndrome can include factors like high blood pressure, excess belly fat, higher than normal triglycerides, high blood sugar and low high-density lipoprotein (HDL) cholesterol (known as "good" cholesterol) (C. Raffaitin et al., Neurology, 2011).[21]

Deepak Chopra – Quantum Healing

Dr. Deepak Chopra is one of the pioneers of mind-body medicine. His name has become a household word to those who are eager to explore alternatives to mind-body medicine, particularly from a spiritual perspective.

One of Dr. Chopra's popular quotes is: "The way you think, the way you behave, the way you eat, can influence your life by 30 to 50 years."

---

[21] https://www.nicabm.com/programs/mindbody/

Another quote that I especially like, since it is aligned with my own beliefs and practices of soulution-based healing: "The less you open your heart to others, the more your heart suffers."

I would reframe Dr Chopra's words and say: "Open your heart and connect with others for a healthy heart filled with peace and joy."

The physical, mental, emotional and spiritual aspects of ourselves are so interconnected, our very lifespan is dependent on loving, life-affirming, soul-connected thoughts, feelings, attitudes, behaviors, outlooks and actions. Moreover, those who are soul-connected, i.e., who have opened their hearts to others (and first to oneself) will have a healthier heart.

Louise Hay, whose bestselling book, *You Can Heal Your Life*, has become a mind-body classic, relates heart disease to issues of the heart. How simple and transparent illness and dis-ease really are![22]

Dr. Chopra was among the first to use the term "quantum healing" when discussing our ability to cure disease from within. According to Dr. Chopra, intelligence exists everywhere in our bodies, in each of our 50 trillion cells, and each cell knows how to heal itself.[23] This resonates with everything Louise Hay, Dr. Candice Pert and millions of others have discovered when they first "opened their hearts" and minds to total body healing and the heart-and-soul connection to our well-being.

Energetics of Healing – Carolyn Myss

The work of Carolyn Myss as a medical intuitive is another important bridge that connects mind-body medicine with the spiritual component. *The Creation of Health,* a book co-authored

---

[22] http://www.amazon.com/You-Can-Heal-Your-Life-ebook/dp/B000SEHQ96/ref=sr_1_1?s=digital-text&ie=UTF8&qid=1449586610&sr=1-1&keywords=louise+hay

[23] http://www.amazon.com/gp/product/B002X4BBK4/ref=dp-kindle-redirect?ie=UTF8&btkr=1

with Dr. C. Norman Shealy, who founded the American Holistic Medical Association, adds credibility to the accurate discoveries of many medical intuitives.[24]

Richard Gerber, M.D., author of *Vibrational Medicine: The #1 Handbook of Subtle-Energy Therapies,* another groundbreaking book in the field of Energy Medicine, has the following comment about *The Creation of Health*: "[This is] an important book that addresses the crucial spiritual issues which lie at the root of many diseases... Shealy and Myss provide a clear understanding of the reasons behind 'dis-ease,' as well as solutions that may heal the higher causes of illness."[25]

In *The Anatomy of Spirit: The Seven Stages of Power and Healing*, Myss describes our "spiritual anatomy" and how its dysfunctions affect the physical body.[26] Myss also presents a complete program for spiritual growth, based on ancient spiritual principles endemic to all of our traditional religions.

One of the most important takeaways from Myss's work is support for the belief that everything pulsates with energy and all of this energy contains information. Through our energy system we are in constant communication with "everyone and everything." All of our experiences, both positive and negative, register as memory in our physical body cell tissues as well as in our subtle energy mental, emotional and spiritual areas.

An intuitive can interrupt the "conversations" or communication that occurs among the different energy fields of the body.

---

[24] http://www.amazon.com/Creation-Health-Emotional-Psychological-Spiritual-ebook/dp/B002PYFVZS/ref=sr_1_1?s=digital-text&ie=UTF8&qid=1449589373&sr=1-1&keywords=the+creation+of+health

[25] http://www.amazon.com/Vibrational-Medicine-Handbook-Subtle-Energy-Therapies-ebook/dp/B005IQ64VK/ref=sr_1_1?s=digital-text&ie=UTF8&qid=1449589262&sr=1-1&keywords=Vibrational+Medicine

[26] http://www.amazon.com/Anatomy-Spirit-Seven-Stages-Healing-ebook/dp/B00DTEMVZG/ref=sr_1_1?s=digital-text&ie=UTF8&qid=1449589570&sr=1-1&keywords=carolyn+myss+books

Medical intuitives serve as energetic receivers who, after interpreting the messages, can transmit healing energy to those areas of weakness that need attention. Often a medical intuitive discovers that the issue causing energy weakness originates in the spiritual part of the energy system. It is a soul issue, for which a soul-ution is required in order to bring the body into balance again.

In *Anatomy of Spirit*, Myss refers to comments made by Dr. Candace Pert, who discovered The Emotion Molecule and whose work we will discuss at length in Chapter Four. When Dr. Pert was a guest on *The Bill Moyers Show*, she stated that we can no longer separate the mind from the body because the same kind of cells that manufacture and receive emotional chemistry in the brain are present throughout the body.

"Sometimes the body responds emotionally even before the brain has registered a problem," Pert claimed. "Clearly, there is another form of energy we have not yet understood. For example, there is a form of energy that appears to leave the body when the body dies....Your mind is in every cell of your body."[27]

Our emotional energy converts to biological energy through a highly complex process. Just as radio stations operate according to designated energy wavelengths, each organ and system in the body is calibrated to absorb and process specific emotional and psychological energies. Each area of the body vibrates and transmits energy at a designated frequency. When we are healthy, every part of the body is resonating at its normal or signature frequency. Like a violin whose strings have been adjusted to vibrate at their signature frequencies, in a state of health, all parts of the body are tuned-up or performing in tune.

By locating the part of the body that is not vibrating at its normal frequency, we can pinpoint the area that needs attention. A change in intensity of the frequency will also indicate a change

---

[27] Op. Cit., Myss, Carolyn, *Anatomy of Spirit*, p. 35.

in the nature and seriousness of the illness. It will reveal the stress pattern that has contributed to the development of the illness.

In learning to tune in to our own energy, we can begin to sense those people, places and situations that increase or deplete our energy.

Laughter, One of the Mind-Body's Greatest Antidotes

By default, Norman Cousins, American political journalist, author, professor and world peace advocate, discovered that laughter is a state of mind that is life-affirming and healing.

In 1964, Cousins, diagnosed with Ankylosing Spondylitis, a rare disease of the connective tissue, was given only a few months to live, with only a 1 in 500 chance of survival. By profession Cousins was a meticulous researcher so he turned a deaf ear on the doctor's prognosis and set to work seeking his own solution. He soon discovered that his disease and the medications were depleting his body of vitamin C.

After firing his doctor, he found another professional who would allow him to be in charge of his own healing. He checked into a hotel, ordered injections of massive doses of vitamin C, obtained a movie projector (this was before videos had been invented) and watched as many funny movies as he could get his hands on. *The Marx Brothers, Candid Camera, Charlie Chaplin* and other favorites were included in his repertoire.

Cousins then proceeded to spend hours watching these films and laughing. Even though he was experiencing much pain, he forced himself to laugh until his sides hurt.

Cousins lived much longer than doctors had predicted—10 years after his first heart attack, 16 years after his collagen illness and 16 years after he was first diagnosed with heart disease.[28]

Questions & Reflections

1. How would you describe psychoneuroimmunology to someone who has never heard of the term?

2. How does this new branch of medicine relate to the statement that "we create our own reality"?

3. What is the difference between Western or allopathic medicine and energy medicine? Why is this important information for you to know, if you are working on clearing your energy and "getting clear" about your purpose in life?

4. If a medical intuitive can shift energy in your body, how can you assist in this process as you learn to sense what your energy body wants?

5. What emotions are stored in your body that could interfere with your own clear energy flow? Fear, pain, and anger can block energy, for example. Can you think of others?

6. If you have non-useful energy stored in your body, where is it located? Use your "inside eyes" and trust what you see.

7. From a physical perspective, why do you think laughter is such a great healer?

Exercise

Take your time and move slowly and purposefully through the following exercise.

---

[28] https://healdove.com/alternative-medicine/normancousins; Cousins, Norman. *Anatomy of an Illness: As Perceived by the Patient.* New York: W. W. Norton & Company, 2005.

You may begin by focusing on your breath. Draw all of your attention to the air you are inhaling and then to the release in exhalation. Continue for several minutes with slow and patterned breathing until you feel quiet and still. As you continue to focus on your breath, further quiet your body and still your mind.

You might sense your feet on the ground and allow yourself to feel your connection to the earth. Imagine that from your feet you can sense the heartbeat of the earth gently pulsating upward through the bottoms of your feet. Feel the energy move up through your feet and allow it to continue to move upward through the crown of your head.

Allow the gentle pulsing to continue to move through you. Notice the rhythmic moving energy clearing any areas in your body. Allow the earth energy to fully and freely fill you.

Use your inner vision to note any areas of potential blockage or slowing down of the flow. In your mind if you see any impediment of energy, ask your subconscious mind to show you a picture or symbol giving you information about that impediment.

As you acknowledge any problem areas, see if you can gently clear them by drawing pure energy of love and support from Mother Earth.

You might want to take notes after your meditation and seek further clearing and balancing of your energy.

You can also pay attention to your dreams. The dream state is subconscious and sometimes even superconscious. In those moments before you drift into sleep and as you are waking up, you are very open and receptive to suggestion, just as in the state of hypnosis.

As you are drifting to sleep you can give yourself the suggestion: "Tonight, if there is anything important related to my healing, it will come to me in my dreams." Or: "Tonight, if there is anything significant that will benefit my soul, show it to me in my dreams." Immediately as you awaken, note the details of any dreams. Write them down. As you practice recalling and recording your dreams, the memories will become more accessible.

In subsequent chapters, I will further discuss the dynamics of subtle energy healing. I will also deliver more information about the science behind our cellular anatomy and the communication network or "world wide web" that exists among our body systems.

As you open your mind and listen to your body, trust that you will always get the information you need.

Visualization

As I mentioned earlier, we "see" only what our mind is primed for. Have you ever looked at an anatomy book and studied the placement of the organs? That is probably the easiest way to help you expand your awareness to understand what your inner eyes see or your intuition senses within your body.

When you are familiar with the structure and function of the energy body, you will be able to more clearly understand the information you receive.

The more expanded our conscious awareness—the more primed our brain is—the easier it will be for us to see and understand specific energy blocks with our inner eyes.

Is it necessary to know exactly what our human parts look like to facilitate healing? NO! In fact, that is one of the most fascinating aspects of our mind-body awareness in regard to healing. You do not need to know exactly what a white blood cell looks like or specifically how it functions. Your body can clearly read your intention and respond perfectly.

The story of Suzette, the nurse who wanted to go on a cruise, is a perfect example of the power of visualization. As a nurse, Suzette had better than average understanding of white blood cells, but it was the visualization we used to boost her immunity that caused her white blood cells to increase. I had her imagine that her white blood cells became a protective army calling for back-up forces.

Suzette also visualized her white blood cells as little "PacMen." She gave them bright headlights and big mouths to see what they were able to "gobble up" in the blood. We suggested not only augmenting the number of white blood cells, but also increasing the current white blood cells' effectiveness in doing their job.

The clear directions you give to your body can be equally effective, barring any emotional blocks you may have to healing.

Exercise

Begin with attention to your breath. Remember, focused attention with a relaxed body can lead you into the hypnotic state in which internal communication is effective.

As you slowly breathe with intention, focusing only on the air you are inhaling and exhaling, move yourself into a quiet place of internal awareness.

Since a healthy immune system benefits all of us, let's focus on that.

Draw your attention into your heart. Imagine as you open your inside eyes, you can actually see your heart efficiently and effectively supporting your body with healthy adequate blood flow.

Imagine that your inner eyes can actually see the blood as if it were under a microscope. Notice the dark red healthy cells smoothly and easily sliding through your blood vessels. You might see them shaped as saucers.

Now notice the white blood cells. Watch them do their perfect work moving through the bloodstream, seeking to find anything that does not benefit your body. As a white blood cell discovers a foreign invader, it immediately vaporizes it!

Watch this process and thank your body for the amazing job it is doing caring for and protecting you.

Gratitude is a self-replenishing attitude your body welcomes.

3

Hypnosis – Healing & Re-Soulution

Hypnosis and hypnotherapy can help us tap into that authentic self or soul to better understand and expand it.

"To thine own self be true," said Shakespeare. Could this not be expanded to say, "To thine own soul be true?"

Subconscious Healing

The unconscious part of the mind-body seems all-knowing and all-powerful. In some therapies, it can be the catalyst for healing or change without the conscious mind ever figuring out what happened.

In 1985 when getting my Master's degree, I was in a small group counseling class with a professor who used hypnosis and regression techniques exclusively. Since we were all in training, during class sessions most of us volunteered to be subjects. I witnessed powerful and shocking insights and healings with my classmates as well as myself.

I was profoundly affected when I discovered that an issue I volunteered in class to work on, namely that I was upset about leaving crying babies to go to school, merely scratched the surface of the *real issue*. In that first regression session, I brought up deep truths that were the core of the "upset" I had been experiencing. I had previously been totally unaware of my own soul truth until that hypnosis session in the classroom.

My own personal experience of insight and the ensuing healing convinced me about the effectiveness of hypnosis and regression. I was equally impressed by the work that I witnessed with the other students. At that moment, I knew that hypnosis and regression techniques would be my path of study. They were a link and a key to what I like to term "express healing."

Little did I realize at that time how profoundly hypnosis would impact my awareness regarding myself and others as I began to use it in my practice. My clients have been great teachers, providing me with a window to see into circumstances and healings beyond my own life experience.

Ultimately in working with the subconscious and superconscious mind, I arrived where I am today with a clear perspective that **we are all souls learning in a human body** and that **everything we choose in life—from careers, relationships, health issues, etc., to each simple thought—profoundly affects us at a soul level.**

With this clarity and insight, my mission and passion are to share with others what I have learned, in order to enable all of us to live with expanded perspective and purpose.

The state of hypnosis and hypnotherapy are the main tools I use with my clients on a daily basis. We can all learn to use these tools to reconnect with our perfect heart. In a subconscious or superconscious state, we can reach within ourselves for guidance and answers. In this state, it seems to be easier to connect with those who have passed, and also to become aware of our eternal nature.

What is Hypnosis?

Many clients come to me after they have been through exhaustive conventional therapy. They don't feel any better even though they've talked out the problem, are aware of their traumas and have made behavior changes. When they are ready to try something different, hypnosis is a great option. I let my clients know that the type of work I do is results oriented and that I will move them as quickly as they can safely be moved. I also tell my clients I do not go on fishing trips looking for issues. Their current feelings will be a great guide to direct the therapy.

Throughout this book, I use the terms "meditation" and "hypnosis" interchangeably since meditation can lead to the state of hypnosis. By quieting and focusing our mind and body we may gain access to previously unanswered questions. Valuable stored information can more easily bubble to the surface of our mind.

Hypnosis is a state of highly focused concentration. It is actually a natural state. Most people do not recognize that they go in and out of hypnosis or "trance," another word for hypnosis, on a fairly regular basis. This state is very safe. In hypnosis, you are always in complete control of what you do and say. In hypnosis, you will never accept a suggestion that doesn't feel right to you and you will never do or say anything against your moral standards.

The person in hypnosis is always in control. No one stays in this state permanently. The hypnotherapist or individual ends the trance formally.

If you have ever been driving on the expressway and suddenly you see the exit sign for your destination yet you have no recollection of having traversed the miles that brought you there, you have been in a state of hypnosis. Your subconscious mind guided you while your conscious rational mind went elsewhere. If you would have encountered an emergency while you were driving, your subconscious would have alerted you, maybe even stepping on the brakes and swerving to avoid a potential accident before your conscious mind ever became aware of the problem.

If you are highly focused on a task or blocking out noises and other distractions, you are in a light state of hypnosis.

Hypnosis is not a sleep state. In the hypnotic state we can be aware of both the conscious and subconscious minds in varying degrees at the same time. Most people will remember everything about this two-way communication during their hypnosis session.

Deeper states of hypnosis are required for pain control. Individuals who are allergic to anesthetics often use a state of deeper hypnosis for anesthesia. This level of hypnosis requires time, training and practice. Woman use hypnosis regularly for pharmaceutical-free childbirth.

During World War II, hypnosis was used as a substitute for chemical anesthesia and for pain relief. In earlier wars and also during the Korean War, psychiatrists and hypnotherapists used hypnosis to treat Post Traumatic Stress Disorder (PTSD). Today it is used in dentistry, dermatology and many other areas of medicine for pain management.[29]

In most therapeutic work a nice light to medium trance state works perfectly. In this state you will still remain aware of what is going on around you. This feeling can be compared to the state you feel when you are about to fall asleep. You know you are awake but you feel so drowsy and comfortable, you don't want to move.

The subconscious mind is the greatest part of our intelligence and is much more expansive in awareness than our conscious, rational mind. Since we act and feel from this

---

[29] http://www.mnhypnotherapyassociates.com/index.php/history-of-hypnosis/

part of our mind, it is the area that is most conducive to healing. Our subconscious awareness can move beyond logic, space and even time. It can remember everything from any time. This awareness allows us to access a place of inner wisdom where deep truths can be brought up to our conscious awareness for the purpose of healing.

Hypnosis is not a zombie state. You will not act like a chicken or bark like a dog unless you decide that is what you want to do. If you have watched stage hypnosis you may have seen people in hypnosis doing some strange things. When you understand why they are doing these things, it will put your mind at ease.

First, in stage hypnosis, the hypnotist will ask for volunteers, usually about 10 of them, to be part of the stage show. The hypnotist will then conduct some relaxation techniques with these candidates to choose which are more responsive to being hypnotized and following suggestions. Often they will send a few people back to the audience if they prove to be unresponsive. About 10-15% of the general population are very suggestible. It is easy for the hypnotist to recognize who they are, so they will be the ones chosen to create the show. These persons would not do anything on stage against their morals or desires.

In the state of hypnosis, the conscious mind becomes aware of the information that surfaces from the subconscious. This is why, during a trance or hypnotic state, a client can continue to carry on a conversation and answer questions.

People are often surprised by the depth of the knowledge and truths they can access in the state of hypnosis. A skillful hypnotherapist can regress a client back to childhood, birth or even beyond, to past lives if necessary.

How Hypnosis Works

The objective when meditating or during a hypnosis session is to access the subconscious mind. Often scientists and artists will find their creative genius surfacing as they quiet and focus, allowing the information that is within to move up into their awareness. Quieting the mind and focusing on the connection of love we have for our departed loved ones can also help us connect with them and recognize the messages they're giving us.

As a note of caution, it is important to be aware that access to the subconscious mind can occur if a person is frightened or traumatized, or under anesthesia.

Once the message is delivered to the subconscious we will act on it automatically. In the case of a trauma, such as an auto accident, our minds can go into what we term "a state of overload or shock." In fact, many of our paramedics and emergency responders are trained in Emergency Medical Hypnosis because it is so powerful. The rescuer may walk up to a victim of a car accident or someone who has fallen off a cliff and knowing they are in an open subconscious state, they will use it to the patient's benefit. The paramedic may say something to the victim like, "The worst is over"; "Help is on the way"; "I am going to put this blanket on you and you will feel more comfortable"; and "Now just slow down your breathing and stop the bleeding from your foot."

The victim responds to the suggestions because in the state of overload or shock, the critical factor of the conscious mind is overridden and the suggestion sinks in. The individual does feel more comfortable when the blanket is placed on them. It is not a magic blanket, but a well-timed suggestion.

A short time ago, a 12-year-old girl came to me for help regarding the release of a recent traumatic experience. When I explained overload and what the paramedics may be trained to say, she said excitedly. "That is exactly what the paramedics said to me and that blanket DID make me feel better!"

Having an awareness of how messages can sink into our subconscious mind is important. When we feel overloaded, we can use these situations in a positive way to drop desirable messages into the subconscious mind. Yet it is often during the times of overload that we pick up messages that do not serve us. We carry some of those messages with us and act on them until they are addressed, eliminated or reframed.

An example of this might be someone in a domestic violence situation. Maybe an individual is being bullied, intimidated and told negative things about themselves. They may begin to believe the violence is their fault, that they are helpless and hopeless, or that they're stuck and must keep trying. Overload techniques are used for brainwashing. This is obviously a block to our soul growth.

I've also seen a few clients over the years who were so frightened in their doctor's office, they left the appointment manifesting symptoms of a disease and suffering from it thereafter. The symptoms of the disease vanished as soon as we removed the suggestion.

First Visit

During a client's initial visit, I take a history of the complaint or issue. I ask a lot of questions and I look at the physical, emotional and soul aspects of the concern. I explain to the client how the mind works and then I teach them self-hypnosis. I record part of the first session so my client can go home and practice retraining themselves in what I label the "relaxation response." In this hypnosis/meditation session I use suggestions to counter what they perceive is the problem.

I want my clients to have relaxation tools in place so they can immediately move into a state of calm at any time, in any place, or in any situation. I am also aware that in meditating they will have better access to the feelings underlying their issues of concern. Should a disturbing memory arise, I want them to know they have effective tools to handle it. Meditation and accessing the subconscious mind also prime us to listen to the messages our mind-body is giving us and to trust the information we're getting.

Sometimes with certain issues—an addiction such as smoking, for example—a single session is all that is needed. The craving vanishes like magic. A few days after they stop smoking, some people find it hard to believe they ever smoked! I've heard the comment more than once that it was just like I had a magic wand.

I do not have a magic wand, but those individuals did not have any emotional blocks. During our session, they merely embraced all the suggestions. At once they "felt" like a non-smoker, which means their mind thought of things other than cigarettes.

Other individuals with a simple issue like a smoking addiction may actually stop the habit, but if they are even still thinking about it, I know they have a block that needs to be addressed. The subsequent session of hypnotherapy is more of a two-way experience. We work between the conscious and subconscious mind of the client. Sometimes we directly address the cigarette; other times we address the part of self that still wants the habit, or we address the pain, anxiety or discomfort the cigarette is masking. Together we find the block and do whatever is necessary to clear it, so the given suggestions will feel completely natural to the client.

On one rare occasion I had a young woman, a nurse, come in for a session to stop smoking. She had no idea why she would smoke; she dealt with the ravages of tobacco with her own patients! She was embarrassed because as a nurse she didn't want anyone to know she was a smoker. After the session, she walked out the door and lit up a cigarette. Then she immediately called me and said, "I'm not ready to give up cigarettes. I now recognize why I smoke. My mother and I had a cup of coffee and cigarette every morning when she was alive. I am not ready to address the grief over the loss of my mother."

In the state of hypnosis her true soul issue became apparent. She longed to stay connected to her mother and recognized she must make the decision to address her grief in a healthy way. As a nurse, she counseled patients regularly in grief, yet had been consciously avoiding it herself. In that case, the cigarettes had been simply a smokescreen, thus preventing her from clearly addressing her own soul work.

Imagine how different it would have been for her to have recognized when her mother died that she and her mother were still connected. Without a smokescreen, she may have been better able to sense that loving connection. A focus on the loving connection that still exists after the death of the loved one can ease the pain.

It is fascinating to note that for most issues a nice light to medium trance in hypnosis is all that is needed. Anyone can access this state. Unlike what is portrayed on television about hypnosis, the client is still able to speak from their subconscious mind, draw in conscious awareness and even become an observer of the process. They are very much in control and actively participate.

Even in a guided meditation for relaxation or sleep only, you may find your subconscious awareness heightened. During the meditation and afterward, you may experience bubbles or bits of wisdom and guidance that you had not intentionally sought. Just take that knowledge and be grateful that the deepest and wisest part of your mind knows what you need. Listen to the messages you receive, especially after meditation.

I encourage all of my clients to meditate daily. It keeps them connected with their heart. They can begin with the guided meditations I give them. Equally important to daily meditation in my mind is for each of us regularly throughout the day to stop what we're doing and breathe with intention. We can take even as little as one minute to quiet our mind, become present in the moment and release our tension.

The power of periodically stopping and breathing cannot be underestimated. Some major companies, such as Google, are actually training their employees in these mindfulness techniques. They find stress levels are reduced, employees are more on task, and it is easier to manage priorities.[30]

As you practice meditation or even mindfulness with your breath, you can greatly reduce your stress level. Since we respond physiologically to our dreams and hold tension in our body, I suggest that as soon as you awaken while still in bed, take a moment to focus on your breath and release any physical tension from your body. You may be surprised that with regular practice you will sense the shift toward feeling more relaxed. As you release tension first thing in the morning and then periodically throughout the day, it can benefit you by not only feeling less stress but also by helping you keep your mind clear and allowing you to stay tuned in to your heart—your internal PGS.

Recently I read that Dr. Chopra also suggests these mini-balancing and breathing sessions throughout the day.[31]

> *At this moment, you are seamlessly flowing with the cosmos. There is no difference between your breathing and the breathing of the rain forest, between your bloodstream and the world's rivers, between your bones and the chalk cliffs of Dover.*[32]

> —Deepak Chopra

---

[30] http://www.fastcompany.com/3053048/lessons-learned/how-google-and-other-companies-help-employees-burn-off-stress-in-unique-ways, http://www.nytimes.com/2012/04/29/technology/google-course-asks-employees-to-take-a-deep-breath.html?pagewanted=all&_r=0

[31] http://www.chopra.com/online-programs/breathwork, http://www.chopra.com/online-programs/breathe-to-balance#sthash.YitKfccJ.dpuf

[32] http://www.chopra.com/online-programs/breathe-to-balance

Regression Therapy

In a state of hypnosis, it is common to use what are called regression techniques. All therapeutic work in hypnosis is always done for healing in the here and now. Regression therapy is an act of going back to an earlier time to retrieve memories that may be negatively influencing our current behavior. This may be the source of an issue that links to the client's current symptoms. In retrieving those past memories and releasing the stored emotion, we can change the way we respond to them.

Regression therapy may be compared to bringing up a past memory and reacting to it in the present moment as if it were a three-dimensional living color experience. In hypnosis we remove the emotion and add expanded perspective and loving compassion so the trauma of the past becomes like an old faded black-and-white photo. We may remember the incident but it no longer evokes emotion.

Some therapists exclusively use regression and past life regressive techniques to heal current issues. Other therapists access in-between life memories. It may even be possible to look into the future to see where you will be, depending on the choices you make now.

When a person begins to meditate, emotions bubble up into their consciousness, or memories of the past can begin to surface. These memories have been buried and may be felt merely as anxiety, but now the mind is quiet enough to recognize the feeling for what it really is.

For deeper levels of hypnosis and regressions, particularly with trauma, it can help to have an experienced guide to keep you focused and moving through the process. When regressing to some painful memories, a guide can also help you observe the situation without physically having to re-experience the pain. Even a person experienced in meditation can benefit from the help of a guide at times.

Throughout this book, I have included guided meditations you can record for yourselves or use with a friend.

At times when I personally want to drop into deeper hypnosis in order to seek information, I will listen to one of my own recordings. This keeps me focused so I will not drop too deeply and can stay on track and remember what I retrieve.

Techniques to regress to a childhood trauma, an earlier time or a past life, are essentially the same. The state of hypnosis may merely be deeper. It is an amazing experience to look into our soul self through our subconscious/superconscious.

When I began this work over 20 years ago I didn't give much thought to past lives. It was my clients who first opened that door of awareness for me, with spontaneous regressions to a different life that was directly related to the core of a presenting issue. After my mind became primed to understand that even memories from long ago could be accessed, I personally experienced many of these memories myself. I was amazed at how healing and mind expanding an experience this can be.

Our earlier experiences retrieved from within ourselves can be reframed or remembered differently, with more insight and compassion. When released, the result can be immediate peace or healing.

If the origin of an issue goes back to a past life, that previous time sheds more light on the current situation. Gaining awareness of the patterns in our life can help us see current circumstances more clearly, and facilitate physical and emotional healing at the level of our soul awareness. It is important in all regressions to learn the lessons from the past so we can evolve.[33]

---

[33] For more information on past life and between life regression, Brian Weiss, M.D. and Michael Newton, Ph.D., have written several books on the subject (See Bibliography at the end of this book.)

I find most of my clients are able to heal from their issues of anxiety, fear, pain and insomnia with hypnosis and regression that focuses on this lifetime. Reframing traumas or perceived traumas usually resolves most of my clients' issues. Re-birthing techniques are also vastly helpful to clear any birth traumas. There are some exceptions.

'I think he's going to kill me!'

"Elizabeth," a forty-year-old nurse, was confused and afraid of her ex-husband, "John." She told me she believed John was capable of murdering her. I asked her why she believed that. She said many years ago, in the early years of marriage, he mentioned to her that he was writing a novel in which she was mysteriously murdered. He asked her at that time to help him figure out whether it would be better to inject insulin under her fingernails or find another natural substance someone could kill her with so no one would discover the cause of death. In the novel, John was to be the hero and figure it out. He never finished the book or brought it up again.

I silently thought, maybe she should have left him then and not waited a decade longer!

Elizabeth said she knew she was highly sensitive. She felt like she could look into people's hearts and see the beauty in all individuals. She loved this aspect of herself and for the most part, it worked well with her patients in the hospital. She couldn't understand the feeling that her ex-husband was capable of murdering her. She was scared and constantly on alert, fearing for her life.

Elizabeth told me she had experienced an unusual incident with John before they were married. In an unexpected moment she looked into his heart and saw him as if he were a perfect spirit body. It was the most magnificent sight she had ever seen and this is what convinced her to marry him, despite a few other concerns.

She reported that after they married, John immediately became distant, accusing, critical and aggressive. Elizabeth said she continued to look into her husband's heart and saw the wounded child. She overlooked the bad behavior and made excuses for him. She thought if she could just be better he would feel her unconditional love.

Eventually, fearing John's increasing violence, she asked for a separation. At that point, John declared all-out war against her. It became a heated battle, increasing her confusion about the perfect heart she saw in him and triggering her fear of his physically harming her.

She said John was a well-respected community leader and logically it seemed highly unlikely he could hate her enough to murder her. People who knew them both asked her what she did to make him so angry. She knew at a deep level that she was not the subject of John's anger even though it was clearly directed at her.

Elizabeth came to me because the fear of being murdered was dominating her thoughts. She took self-defense training, changed the house locks and began to lock her car. She didn't want to live in fear anymore. I could see that Elizabeth really believed what she told me.

We used hypnosis first in a guided meditation to help her relax, sleep better, and focus on her own healing and grief. She handled the stress better but the fear still remained. I regressed Elizabeth to the first time when she felt the profound fear that her life was threatened. She went immediately back to a past life somewhere in Northern Europe. She described a scene in detail: the tall rock walls of the castle where she was standing. A gentle breeze riffled through her long blonde hair, her filmy dress billowing around her.

Elizabeth was aware that she came from a royal family with power. She saw a young man near the castle wall, clearly from the lower class, whom she sensed very much wanted a position of power. She recognized this young man as her former husband in this life. At that time in the past she looked inside of him and saw his great capability as a leader. She knew she could help him achieve his goals. She married him but he resented her for her beauty and power and he killed her shortly after he got the position he wanted.

Elizabeth opened her eyes and said, "That is exactly what happened in this life. I saw John's poor upbringing and helped him achieve his goals. He has resented me ever since. He did kill me before, but that will not happen again!"

With that awareness, Elizabeth immediately released her fear of John harming her in this life. She understood why he resented her so deeply. It went beyond reason. Elizabeth stopped worrying about her life being threatened. She also recognized she had learned a great soul lesson of love in her relationship with John. She had learned to love someone who was unkind to her. She now felt compassion for John. She also learned that just seeing a beautiful heart is not sufficient reason for choosing that person as a life partner. She learned that she must balance her heart feelings with her rational mind.

There are some who will say that memories of a past life are simply a metaphor for a current situation. Nevertheless, metaphor or past memory, Elizabeth's fear completely dissipated.

Rebirthing

I have also developed a way to regress a person back to their perfect soul state just before they enter their body so I can add energetic protection to the developing child. This is particularly helpful when an individual had a strongly emotional mother during their gestation period. I lead those souls in a visualization of a beautiful shield made of colored light that wraps around the fetus to protect the baby from picking up parental fears or strong emotions. Those strong, especially maternal emotions are the beginning of the neural circuitry in the developing baby. It is amazing to witness the profound healing that occurs by clearing early traumas through rebirthing techniques.

The individual then can imagine being reborn to enlightened loving parents. If the parents were not damaged from their own traumas and thus could see their baby as a perfect soul child, they would be loving and caring. We reframe not only the actual birth but also the gestation period.

Gestalt Dialogue

Using what is called Gestalt Dialogue in hypnosis can be very enlightening. Gestalt work results in increased and enriched awareness and expanded perspective. Often, this leads to feelings of compassion for others who intentionally or unintentionally hurt us. The understanding that occurs during this process brings resolution, growth and peace.

I used this type of hypnosis with a client named "Gina," who presented with anxiety and fear of intimacy resulting from what she described as her alcoholic mother's rages. She said her father died when she was young and she didn't remember anything about him.

My Parents Love Me

In hypnosis, we brought up Gina's feelings of anxiety and traced them back to three-year-old Gina traumatized by her mother's raging behavior immediately after her father died. Gina then remembered her father lying in bed when he was sick, and felt her confusion at his unexpected death. The young Gina wondered, "Why did he abandon me by dying?" Not only had she lost her daddy, which she had never acknowledged until returning in hypnosis to that time, but she also realized that she thought her mother's anger was directed at her, that somehow she was responsible for her father going away. Thus, she protected herself from becoming emotionally close to others.

In hypnosis, Gina spoke to her father, whom we put in a chair in front of her. She expressed to him her love, confusion, anger and feelings of abandonment. I then had Gina speak and become her father's voice. In her voice, she spoke his words and feelings back to her. He told her calmly and slowly that he loved her very much and that it was simply his time to die. He assured her he only felt love for her and had continued to watch out for her.

This was powerful and moving for Gina. Until that moment, Gina had not recognized the depth of love she had for her father and that he also had for her. She could then begin her grieving process for him.

Next we put Gina's mother in the chair across from Gina. Three-year-old Gina began by saying, "Mother, I want to tell you how I feel when you scream at me." Gina went on to describe the feeling that her mother hated her, that she could never do or be enough to make her mother happy, that she lived her entire life in fear of others and that she was angry at how her mother had treated her.

I then asked Gina to become the voice of her mother, to talk back to the three-year-old.

Gina then traded places in the chair with her now deceased mother and became her mother's voice. She was surprised at the pain and confusion her mother felt. Her mother was totally overwhelmed and angry at God about losing her husband. She assured Gina that none of her angry behavior was Gina's fault. She was scared, shocked and depressed, and she felt abandoned by her husband. Gina could then see right through her mother's behavior into her broken heart and could look at her with compassion. Between the two of them (in dialogue in the chair) they came to a place of resolution.

I then asked Gina if there was anything else she wanted to tell her mother. In a soft, three-year-old broken voice she said a heartfelt, "I love you, Mama." I asked if her mother had anything to say to Gina. Mother replied: "I love you too, my precious daughter." Gina was able to feel worthy of love and gained compassion and understanding of herself and her parents.

We then brought Gina's father back and sat him next to her mother. She could see the two of them holding hands and realized her parents had loved each other very much.

I asked her if she could feel what it would feel like for three-year-old Gina to be embraced in a group hug by both her mother and father. With great emotion, she said she could feel the love embracing her. I could also feel the love of that family unit in the room.

I pointed out that these soul relationships were eternal. She could look to her parents any time and feel only their love. I saw her breathe a sigh of relief and a deep look of peace washed over her face.

Gina was able to expand her perspective by looking and feeling through her parents. She could now view her childhood with heightened awareness and gain understanding from her parents' viewpoint. The reality was that Gina was not abandoned; she was loved. Her mother simply didn't have coping skills.

We then had Gina imagine what her life would have been like with both mother and father present in a loving supportive relationship. Clearly they would have provided this for her if that had been possible.

As Gina was able to shift her perception of childhood, she began to love herself and recognize she was worthy of love, and that she was lovable.

As we ended our session and Gina wrote a check, she looked at the date on the check. She took a deep breath. Wide-eyed with surprise, she said, "Today Is January 22. That is my father's birthday!" The synchronicity was a lovely gift of forgiveness and understanding for all three of them.

As we do our own work clearing and understanding our past issues, we reach a place of forgiveness, empathy and understanding of ourselves and others. All previous experience becomes a base for learning and soul growth.

Empathy for others is a key to understanding and forgiving them and allowing us to feel compassion. As individuals, we have the power to positively influence our own future lives. As humans collectively, our choices influence our collective future.

Gina needed to learn not only to give and be loving but, equally important, she also needed to allow herself to receive love. This is a valuable life lesson for all of us.

Hypnosis Can Remove Blocks

As we discussed previously, during our growing years and even as our body is developing before our birth, our subconscious mind retains feeling-data from our environment that forms the basis for the programs from which the ego or conscious mind operates thereafter. It is possible that these early programs become blocks that energetically stand in the way of allowing us to achieve our full potential.

Working in a state of hypnosis is one of the most effective ways to help people identify and release energetic blocks so they can reconnect with their soul and continue their soul journey.

Hypnosis Can Erase Family Histories of Illness

During our sessions, "Robert" related a multitude of lifelong traumas and problems, including a strong family history of mental illness. He suffered from anxiety and had been hospitalized for mental breakdowns as well as other health challenges.

In the first couple of sessions, Robert was able to relieve some of his current physical symptoms. He was an excellent hypnosis subject, easily moving deeply into trance. I moved him through remembering the birth process. He was surprised to find his parents expected him to be gravely deformed because his mother had been on powerful drugs during the pregnancy. He remembered his father counting his fingers and toes at birth. They were looking for something wrong with him and that message sank into his mind.

I took Robert back into the light just before he came into his body and we gave him a beautiful protective shield of colorful light that we wrapped around him during the gestation period. When he was born he was welcomed by parents delighting in his perfect health as they counted his fingers and toes. Robert immediately shifted his awareness and started to recognize his tremendous strengths. His health greatly improved.

For some individuals, an initial trauma could have occurred in a previous lifetime. In these cases, a past life regression is beneficial. A look into the past can often help us heal the present.

Individuals can experience hypnosis and regression in different ways. A number of techniques focus on a strong emotional response in order to access and reframe memories. It is also possible to move back in time and be an observer of circumstances. Some people see scenarios as if they were watching a play. Others have very detailed memories. They can recall previous places, clothing, even shoes in detail. Still others may have strong sensations, such as hearing, smelling, tasting or touching, that are related to these events.

I Want to Be Healthy

Many years ago, "Gladys," a woman in her seventies, came to me for smoking cessation. She had started smoking in her thirties when she was diagnosed with post-polio syndrome resulting from a previous bout with polio. At the time, in the late 1940s, she was a newly divorced single mother who was experiencing physical pain.

The doctor who diagnosed the post-polio syndrome told her she would have pain every day for the rest of her life. He said she could take pain medication daily or she could start smoking. She was told the nicotine in the cigarettes would numb her nerve endings and help alleviate the pain. Gladys opted to smoke, although she detested cigarettes.

Each morning she would lock herself in the bathroom and smoke several cigarettes so she could feel better. This went on for almost forty years.

Gladys came to me for help in learning a better way to deal with the pain. We eliminated the cigarettes in a single session. I followed up the next week by tracing her pain back to the time in the late 1940s when she was divorced and became a single parent. Her painful emotions manifested as physical pain in her body. In one session we reframed her emotional pain from the divorce. Her physical pain completely disappeared. She was shocked. I seriously doubt that she ever had post-polio syndrome, but her body believed she did and she suffered for over 40 years.

Gladys then had another problem. She asked, "How do I tell the government? They have been paying my rent for all these years." She was now no longer disabled by pain. I told her that since she was in her seventies and on Social Security, she should probably not even attempt to tell that story and just enjoy the rest of her life in comfort.

That case made a big impression on me. Here was someone who suffered four decades with symptoms of a painful disease that we eliminated in an hour! This woman was frightened and overloaded when the doctor told her she would have pain every day for the rest of her life. This message sank in deeply. Beware of feeling overloaded or overwhelmed anywhere, even in the safety of a doctor's office!

Overload techniques in therapy can be used safely by a well-trained regression therapist to bring up a strong feeling and trace it back to its root cause for healing purposes. In this circumstance, a troubling emotion can be traced back to when it was first experienced in the body, cleared and reframed so the symptoms disappear. In my training, we call this "transforming therapy." It is truly life transforming.

A third way a message can be delivered to the subconscious mind is when a person is in a coma or under anesthesia. Normally, the medical staff is well-trained and keeps conversation neutral in the operating room and ICU. Occasionally, I will have a client present to me in the office with anxiety after a medical procedure. The doctor assures them everything is just fine but the patient has an underlying, unexplainable anxiety with an onset immediately after surgery. In that case, I regress the patient back into the surgery and, believe it or not, they can remember what was said during the surgery or in the recovery room that triggered the problem. We then reframe the message and the anxiety vanishes.

If you are going to have a surgery, you can ask your anesthe-siologist to give you healing suggestions to compound your body's own innate ability to heal. Examples of suggestions are: "Your body is healing quickly and easily"; or: "Every easy breath you take helps you to feel more comfortable and brings your body into perfect balance."

Deep soul healing occurs in a hypnotic state. Sometimes in deep hypnosis or dreams, we can connect with what we call the superconscious state. This is a place of enlightenment and wisdom.

Hypnosis Can Release Anxiety

A while ago I had a woman in the office suffering from debilitating anxiety and obsessive thoughts about a job she quit 25 years previously, to pursue another very successful career. She had no idea why these overwhelming thoughts were dominating her mind at this time.

At the end of the first basic session, as she opened her eyes she said aloud, "I need to retire."

I responded, "That is an important piece of information to consider."

"My husband doesn't want me to. I make too much money," she said. She discovered the cause of her anxiety, a message from her subconscious, that she'd had enough of just pursuing money. She was worn out from the constant pressure.

She then had a choice to continue to ignore her inner wisdom and deal with her anxiety or listen, speak up and feel peace.

The important part of soul work is to help us clear our mind-body of old messages that don't work for us. We can then embrace our truest nature. As we clear out habits, anxieties, stories or rules, fears and addictions, we are moved toward love and compassion not only for ourselves but also for everyone else. We can become more aware of our heart and soul center.

Even in the case of regression back to a childhood trauma, it is possible with expanded perspective to look with compassion and understanding at those who harmed us. Remember, love is our true nature.

Develop a Sense of Knowing

An awareness of knowing that things are coming to you from the outside begins with knowing yourself on the inside. What parts of you are innate traits you had before you were born that are connected with the soul part of you? Which of these were influenced by the chemical soup of your mother's emotions?

In utero, the fetus develops into its own unique combination of genes from each parent that also includes maternal biological emotions as well as external influences on the fetus from the host (mother).

A mother's chemistry is always shifting. When a mother thinks about something that evokes strong fear, at once this emotion triggers the body into a state of high alert, releasing a number of chemicals into the bloodstream. The baby, although a unique individual, is dependent on its mother for life and shares the same blood. It will also feel the chemical results of the mother's fear but not understand what the emotion means.

It has been my experience that a woman's feelings in giving birth to a child may be similar to her own feelings of being born. That experience may retrigger a birth or perceived birth trauma in the mother, resulting in postpartum depression.

Over the past 20 years, of all of the individuals whom I have helped prepare for childbirth, only one person has had postpartum depression. That exception was a physician who refused to deal with her strong negative issues regarding her mother.

The process of sensing in our body begins before we are born and continues throughout our life. Our conditioned thought process combined with subconscious programming may be what prevents us from listening to our deeper selves.

Reconnecting with that deeper, wiser and ever evolving self by removing the layers of conditioning and insulation can put us back on the path we were born to walk, using our body and minds to sense, feel and trust the information we receive.

Knowing requires clarity. You can trust the truth of the information you receive through your thoughts, feelings and moments of synchronicity when you walk your walk and keep your body clean and clear.

In general, the cleaner our bodies, the more astute are our minds. Pure foods and water, clean air, and avoidance of chemicals can sharpen our sensitivities and help our physical bodies experience accurate feelings so we can trust the information we receive.

Questions & Reflections

1. Do you use self-hypnosis? If so, how do you use it? Is it effective? Explain.

2. Is meditation a part of your lifestyle? Can you identify how it has enhanced your life?

3. Have you used hypnosis for resolving an issue? Was it successful, i.e., did it work well for you? Explain.

4. Have you ever been regressed? Did you learn new things about yourself? Was this information helpful? Explain.

4

The Power of Our Emotions from a Scientific Perspective

I've come to believe that virtually all illness, if not psychosomatic in foundation, has a definite psychosomatic component.

—Dr. Candice Pert, Neuroscientist

In 1884, when he was an assistant professor at Harvard, William James published an essay titled "What is an Emotion?"[34] In the essay, he claims that the source of emotion is purely visceral, that it originates in the body and not in the mind. Little did James realize that he had tapped into a whole new way of viewing the body systems. We've come a long way since then.

Neuroscientist Dr. Candace Pert's breakthrough discovery of The Emotion Molecule solved another mystery regarding the way the body receives, processes, stores and transmits information.

The immune system is an integral part of the nervous and endocrine systems. States Dr. Pert in *Molecules of Emotion: The Science Behind Mind-Body Medicine*:

---

[34] You can read the entire essay at http://psychclassics.yorku.ca/James/emotion.htm

> The three classically separated areas of neuroscience, endocrinology, and immunology, with their various organs—the brain; the glands; and the spleen, bone marrow, and lymph nodes—are actually joined to each other in a multidirectional network of communication, linked by information carriers known as neuropeptides.[35]

According to Dr. Pert, it took 15 years of research before scientists dared call the neuropeptide biochemicals "molecules of emotion." Hundreds of scientists mapped the location of these molecules and found them in every cell of the body.

This discovery canceled out previous unproven but scientifically accepted beliefs that the mind and body are separate from each other and the mind controls the body. Dr. Pert reminds us that your body does not exist just to carry your head around! Emotions are a vibration that happens simultaneously throughout the body.

This is important to us because it means that when we change the way we fire up our brain cells and thus the way they are wired, the emotion molecules in every cell of our mind-body instantly deliver this message of change to every part of our being.

States Dr. Pert:

> I've come to believe that virtually all illness, if not psychosomatic in foundation, has a definite psychosomatic component. The "molecules of emotion" run every system in our body creating a bodymind's intelligence that is wise enough to seek wellness without a great deal of high-tech medical intervention.[36]

---

[35] Pert, Candace D., *Molecules of Emotion: The Science Behind Mind Body Medicine.* New York: Simon & Schuster, 1999.

[36] Op. Cit., Pert, Candace, p. 184. http://www.nytimes.com/2013/09/20/science/candace-pert-67-explorer-of-the-brain-dies.html?_r=0

From Dr. Pert's findings, we've learned that every one of our mind-body cells has a molecule of emotion located on the cell's surface. This molecule is called an opiate receptor. Receptors function as sensing molecules or scanners, just as our eyes, ears, nose, tongue, fingers and skin act as sensory organ receptors.

Receptor molecules have "keyholes" that cluster in the cell membrane waiting for the right chemical "keys" or "ligands" to swim up to them. When the right ligand fits into a receptor keyhole, the process is known as binding.

As the ligand enters the receptor, it creates a disturbance that tickles the molecule into rearranging itself and changing its shape. In the process—click!—information enters the cell.

This information can be a feeling, for example. When the ligand binds to the receptor, it transfers a message via the molecular properties to the receptor.

To further illustrate this process, consider two voices—ligand and receptor—striking the same note and producing a vibration that rings a doorbell to open the doorway to the cell. The receptor, having received a message, transmits it from the surface of the cell deep into the cell's interior, where the message can change the state of the cell dramatically.

The message of the ligands initiates and directs a chain reaction of biochemical events, e.g., making decisions about cell division, opening or closing ion channels, and adding or subtracting energetic chemical groups.

The receptors on the surface of the cell determine the movement or non-movement (change) of the cell. This is based on whether ligands occupy the receptors. No ligands entering the receptors: no action, no change. No emotion (e-motion), no chemical or physical change.

If the cell is the engine that drives all life, then the receptors are the buttons on the control panel of that engine, and a specific peptide is the finger that pushes the button and gets things started.

Physical changes at the cellular level can translate into larger mind-body changes in behavior, physical activity, even moods. When a change is "fixed" in the unconscious part of the mind-body, it becomes a program.

We perceive events and have bodily feelings. After the perception joggles our memories and imagination, we name our physical sensations, labeling them as one of our many emotions. The body perceives and the body responds: pounding heart, tight muscles, and sweaty palms.

Our emotions are felt throughout the body as sensations. Each morsel, with its pulsations of feeling, dim or sharp, pleasant, painful or dubious, contributes to the shaping of our personality.[37]

Writes Dr. Pert:

> Emotions are constantly regulating what we experience as "reality." The decisions about what sensory information travels to your brain and what gets filtered out depends on what signals the receptors are receiving from the peptides. There is a plethora of elegant neurophysiological data suggesting that the nervous system is not capable of taking in everything, but can only scan the outer world for material that it is prepared to find by virtue of its wiring hookups, its own internal patterns, and its past experience. The superior colliculus in the midbrain, another nodal point of neuropeptide receptors, controls the muscles that direct the eyeball, and effects [sic] which images are permitted to fall on the retina and hence to be seen.

---

[37] Pert, Candace, PhD. *Molecules of Emotion: Why You Feel the Way You Do.* New York NY: Scribner, 1997. http://www.amazon.com/Molecules-Of-Emotion-Mind-Body-Medicine/dp/0684846349

Peptides weave the bodies [sic] organs and systems into a single web that reacts to both internal and external environmental changes with complex, subtly orchestrated responses.

...God is in the frontal cortex! As the part of the brain that gives us the ability to decide and plan for the future, to make changes, and to exert control over our lives, the frontal cortex seemed to me to be the God within each of us.[38]

Wisdom of the Body – Innate Intelligence

The mind as we experience it is immaterial and yet has a physical substrate, which are both the body and the brain. The mind may also be said to have a non-material, non-physical substrate that monitors or controls the flow of this physical "information." It is the mind that holds together the network, often acting below our consciousness, linking and coordinating major systems and their organs and cells in an intelligently orchestrated symphony of life.

We can refer to the whole system as a psychosomatic information network, linking the psyche, which compromises all that is of an ostensibly nonmaterial nature, such as mind, emotion, and soul to *soma*, which is the material world of molecules, cells and organs.

"We can no longer think of the emotions as having less validity than physical, material substance, but instead, must see them as cellular signals that are involved in the process of translating information into physical reality, literally transforming mind into matter," writes Dr. Pert. "Emotions are at the nexus between matter and mind, going back and forth between the two and influencing both."[39]

---

[38] Op. Cit., Pert, Candace, *Molecules of Emotion: The Science Behind Mind-Body Medicine*, pp. 147, 148.

[39] http://www.wellnesstips.ca/Mind%20and%20Body.htm

When we are nervous, often we feel butterflies in our stomach. The intestinal tract has thousands of neuropeptides with receptors, so it would be natural that we feel our emotions in the gut, having gut instincts, etc.

Endorphins are the "feel-good" hormone. When a woman is in labor she is taught how to consciously alter her breathing, i.e., breathe deeper, in order to reduce pain, since deeper breathing relaxes the body and reduces pain by altering the quantities of endorphins that are released.

Likewise, if you vividly imagine slicing and then biting into a juicy lemon, you will probably find yourself salivating. Your thought created a physical response in your body. Just as your physical body can influence your emotions and your thoughts, your thoughts and emotions can change your physical responses. This happens through peptides being released and binding to receptors as a consequence of your physical actions, your thoughts or your feelings.[40]

As mentioned previously, Dr. Pert's and other scientific findings eventually led to the development of the new field of psychoneuroimmunology, also known as mind-body medicine.

The Brain & Its Relationship to Mind-Body Function

Think of the brain's function as not merely one of filtering and storing sensory input, but also of associating the sensory input with other events or stimuli occurring simultaneously at any synapse or receptor along the way. This is the process of learning. One extremely important purpose of emotions from an evolutionary perspective is to help us decide what to remember and what to forget.

---

[40] http://www.wellnesstips.ca/Mind%20and%20Body.htm

Mind-body therapies such as biofeedback, a technique of using monitoring devices to measure the various bodily functions, e.g., heart rate and blood flow, can help people attain a deep state of relaxation. This makes it possible for them to take conscious control of physiological processes that were previously thought to be autonomic or unsusceptible to voluntary intervention. For example, anyone can increase the temperature of their hands 5-10 degrees on the first try.

Most of our mind-body attentional shifts are subconscious. While neuropeptides are directing our attention to their activities, we are not consciously involved in deciding what gets processed, remembered and learned. However, we do have the possibility of bringing some of these decisions into consciousness, particularly with the help of various types of intentional training.

Through visualization, for example, we can increase the blood flow to a body part and thereby increase the availability of oxygen and nutrients to carry away toxins and nourish cells. Neuropeptides can alter blood flow from one part of the body to another. This is an important aspect of prioritizing and distributing finite resources available to our body.[41]

Conscious Breathing

Conscious breathing is extremely powerful. A wealth of data shows that changes in the rate and depth of breathing produce changes in the quantity and types of peptides that are released from the brainstem, and vice versa.

By either holding your breath or breathing rapidly, you cause peptides to diffuse rapidly throughout the cerebrospinal fluid in an attempt to restore homeostasis, the body's feedback mechanism for restoring balance. Since many of these peptides are endorphins, the body's natural opiates as well as other kinds of pain relieving substances soon diminish your pain.

---

[41] Op. Cit., Pert, Candace. *Molecules of Emotion: The Science Behind Mind-Body Medicine,* p. 147.

The peptide respiratory link is well documented. Virtually any peptide found anywhere else in the body can also be found in the respiratory center. This peptide substrate may provide the scientific rationale for the powerful effects of consciously controlled breathing.

The concept of a network, stressing the inner-connectedness of all systems of the organism, has a variety of paradigm breaking implications. In the popular lexicon, these kinds of connections between the body and the brain have long been referred to as the "power of the mind over the body." That phrase does not describe accurately what is happening, however.

According to Dr. Pert, mind does not dominate body. Mind *becomes* body; body and mind are one. The body is the actual outward manifestation in physical space of the mind:

> "Bodymind," a term first proposed by Dianne Connelly, reflects the understanding derived from Chinese medicine that the body is inseparable from the mind. And when we explore the role that the emotions play in the body as expressed through the neuropeptide molecules it will become clear how emotions can be seen as a key to the understanding of disease.
>
> We know that the immune system like the central nervous system has memory and the capacity to learn. Thus, it could be said that intelligence is located not only in the brain but in cells that are distributed throughout the body and that the traditional separation of mental processes, including emotions from the body is no longer valid. If the mind has been defined as brain cell communication then this model of the mind can now be seen as extending naturally to the entire body. The mind is in the body the same as the mind is in the brain.[42]

This explains how, in the state of hypnosis, we can actually talk to parts of the body and get accurate and exact information.

---

[42] Op. Cit., Pert, Candace, *Molecules of Emotion: The Science Behind Mind-Body Medicine*, p. 88.

> Every second a massive information exchange is occurring in your body so much so that the term mobile brain is an apt description of the psychosomatic network through which intelligent information travels from one system to another.[43]

My clients are often delightfully surprised as I lead them in hypnosis to speak with their body. What is most surprising to them is the information the body imparts to them. Although it is exactly what they need, it may surprise them.

Several techniques may be used for requesting wisdom from the body. One is simply to directly ask the body questions. In the state of hypnosis, I will ask my client to imagine their body is standing or sitting directly in front of them. They then ask the body a specific question. The client becomes the voice of the body and answers this question out loud. We continue to ask questions and receive responses.

As we gather the information needed to heal, we draw up a contract between the client's mind and the body's wisdom negotiating what is needed to give the body what it wants to heal. I actually have both the client and the body imagine signing this contract. We place a specific date on the contract and I say I am stamping it "Complete." We roll up the contract and store it in the client's memory bank.

I cannot tell you exactly how it works but I can assure you if the individual keeps their part of the contract, then the body will keep its part as well.

This might sound a little confusing because we know the mind-body is one. Yet I will tell you I consistently see the healing results from this work.

We can use numerous other indirect ways to seek information for healing. The mind responds well to symbols, for example. In

---

[43] Ibid.

the state of hypnosis, by allowing the mind-body to bring up a symbol of something that is needed for healing or something that needs to be released, we bypass the critical mind. In exploring the symbol, often we find one symbol is worth a thousand words.

Another technique that I find extremely effective is for a client to imagine seeing their body in front of them as an energy being. I will then direct my client to focus on a specific body part and ask what color appears. I usually ask about non-symptomatic parts of the body first.

We then explore the color as to whether or not it feels like a healthy one. If healthy, we leave that body part and move on to another part. If the color that is "seen" with the inner eyes is not healthy, I will regress that person to the time when that body part had a healthy color. In their mind they will arrive at a different time and place. Then we will advance time as if watching a movie one frame at a time and we will observe the color of that body part. As the movie continues, in one of the frames they will see the body part shift away from the healthy color.

Just as the color is switching, I direct them to the situation and emotions that were present at the time their body part went from healthy to less than healthy. It is at that point where we will do whatever emotional work is necessary to clear the trauma, misperception or emotional blockage so the body part can return to its healthy state.

Once the emotional blockage is clear, cellular communication will become more efficient.

With this type of work, you can see that the body, or deep wisdom of the soul, contains all the information we need.

'I Need Fast Healing'

"Julie," a busy young mother, came to my office to prepare for her upcoming surgery. She was aware of the power of her mind

71

and hoped to speed up the healing of her torn ligament. Julie was also a soccer player and she was unhappy about her doctor telling her she would need a full year of recovery before she could play soccer again.

She had just visited her doctor that morning and had an MRI. The doctor told her clearly that her body could not heal without surgery. The surgery was scheduled for the next day.

We made a recording for Julie to listen to during the surgery. In it we used the terminology that the ligament was already re-attached and that she had a construction team carefully rebuilding the cells to form a strong and perfect connection for the previously torn ligament. Julie imagined herself playing soccer with grace and ease.

The next afternoon I received a call from Julie, "You will not believe this," she said. "My doctor opened up my leg in surgery and the ligament had started to reattach on its own. He said my body was already mending and I could be back to normal activity in six weeks!" The doctor told Julie he had just seen the detached ligament the day before and had no explanation for her miraculous healing.

Both Julie and I were surprised. The visualization we had done was meant for the surgery and post-recovery, but her body was ahead of the game. Somehow, according to her doctor, she did the impossible.

What is the relationship of the mind and emotions to a person's state of health?

If the immune system and healing such as occurred for Julie can be altered by conscious intervention, what does this mean in the treatment of other major diseases?

The idea that emotions are linked to cancer has been around for a long time. Back in the '70s when I worked on a cancer ward in the hospital, guided meditations were just beginning to be introduced in the medical world. I noticed some people responded well and others did not. At the time, I didn't understand why.

Since then, we have experienced advances in mind-body wisdom. With a better understanding about the energy we store as unexpressed emotion, viz., that it may interfere with healing, the inconsistencies of the past make more sense. This concurs with what I have found in my healing work. If a person has a subconscious block, such as a deep unresolved conflict they may not be aware of, this block or conflict may interfere with the body's communication network and thus impede the healing process.

In the state of hypnosis, those issues can be discovered, remembered and addressed, thus opening the channels of energy to promote healing.

Mind-Body Wisdom

Years ago I was called by a frantic husband to help his wife in premature labor. He said she was currently at home but she had two hours to stop the labor or she would have to be hospitalized. It was the end of the day so I drove to their home.

"Joanie" was indeed in labor and she was only six months into her pregnancy. It was critical for her to stop labor to maintain that pregnancy. We worked only once, for an hour, and the labor stopped. I made her a recording to listen to daily. When the baby was full term, she gave birth to a healthy baby boy.

Months after the baby was born, I received another frantic call, this time from Joanie. She had just discovered that her husband had been having an affair. Actually, he had been having the affair during her pregnancy, in her own home, when she was on bed rest during the pregnancy. She was as upset as her husband had been the day he called me during her pregnancy when she had started premature labor.

We began to work on her issues of fear and grief, but Joanie moved away before we finished the work. Fast forward two years. Joanie called me once again and informed me she had been diagnosed with breast cancer. Once again, we had only a couple of sessions to do the work, not enough time for her to move through her grief and anger. I did not feel a sense of completion in our work, yet I hoped all would be well for her.

A year later, Joanie called again. The cancer had returned and she was ready to do the work. Twenty years after that time, she lives a beautiful, healthy life and her children are grown.

Looking back, it is easy to see the pattern that began during Joanie's pregnancy. Her wise subconscious mind was aware of something that was very disturbing. On a conscious level, Joanie knew nothing of her husband's infidelity. Her uterus was signaling "danger" to Joanie and her husband that something was seriously out of balance with their marriage. Her body knew something was wrong.

Fortunately, in just one session we overrode the premature labor, so the baby could be born healthy. Yet in the years that followed, Joanie continued to ignore mind-body messages pertaining to the imbalance of her emotions until the return of cancer got her attention. In clearing the previously stored emotion, her body returned to a healthy balance.

Early Warning System

During the pregnancy with my quadruplets, I was also in a highly sensitive, extraordinary intuitive state. For most of the pregnancy I was on bed rest and I could tune in and listen well to my body.

One night I awoke in the middle of the night with a feeling of great dread. I was living in San Diego at the time, at my parents' home. My husband was in the Air Force in New Mexico, where he had just been stationed. After the children were born, I planned to join him there. In preparation, we had purchased a home. It was a new home that was currently being built. I had worked closely with the builder, choosing every last kitchen appliance and doorknob!

Initially, I did not want to buy a home in that area because I knew it was not a place where I would want to permanently live, but I went along with my husband's wishes; he was eager to settle there.

On the night I awakened with dread, I absolutely knew without a doubt that we could not purchase that home. I tried to talk myself out of what I was being told intuitively because I knew it would cause major marital discord. My normal mild Braxton Hicks contractions grew stronger and stronger. I was six months pregnant; those babies were too small to be born. The more I tried to talk myself out of what my body was telling me, the stronger the contractions became.

I called my husband, who was a very logical rational physician, and told him we could not purchase the house. We needed to back out of the contract and lose our deposit. I was grateful that he was 1000 miles away. There was not a single logical reason why a woman who was to give birth to four babies and who already had a two-year-old, would *not* want to have her nest settled.

I knew I had no choice. Either I persisted with my decision not to move into that house or I would lose the four babies. After a "heated discussion," I informed my husband, "We will not buy the house. I refuse to sign the paperwork."

He was shocked. I had never before been so bold.

As soon as I stated my position to my husband, my labor stopped. I had no idea why all of this was going on; it certainly was not convenient.

The babies were born on November 11. I was scheduled to return with the children to New Mexico on Dec 10. Although I did not know it, on December 8, the house we had not purchased and that had been sold to someone else, had flooded due to extreme weather. It was completely unlivable.

How did my body know? Was I prompted by Spirit? Protected by angels? Or was I picking up the energy of a house being currently built that had a flaw? Whatever the explanation, I was deeply grateful. In addition, ultimately I lived in New Mexico for only three weeks. The babies, sensitive to the high altitude, got sick and we had to leave the area.

Had I not listened to my intuition, there would have been no home to go to and it would have been unsellable, creating even more unnecessary financial stress.

Was It Not Safe?

"Polly," 39, came to my office for her fear of public speaking. She was a scientist who worked to protect animals. After two sessions, at her upcoming conference, she was able to speak with confidence and ease. As soon as she felt like her job was to speak on behalf of the animals she loved, her fear of public speaking vanished.

She then mentioned that she was also very distraught at the discovery that her father had molested her niece. She had always had weird feelings about her father and didn't feel safe around him. Now she knew why. Polly had also been trying to become pregnant for over ten years without success. She had given up trying to conceive.

We did the work to clear her early childhood memories, releasing her own trauma regarding her father. Six months later, Polly called to tell me she was pregnant!!! She was shocked and delighted.

Why did her body suddenly become pregnant? Did her body know it wasn't safe to bring a child into the world with her subconscious secret? Did her newly found, yet old knowledge, give her body the sense of safety it needed in order to allow a new child to be added to the family? Nine months later she delivered a perfect child.

My Throat Closes Up

"Monique" presented as a lovely young woman filled with anxiety. She reported that her childhood was kind of "weird" but she had been out of the house for over a decade and didn't think her family had anything to do with her current anxiety. She was in a solid long term relationship of 10 years with a man named Bob who she said was very good to her and whom she dearly loved.

She recognized that in spite of Bob's "perfect" track record, often she felt extremely anxious if he was even a few minutes late coming home from work. She recognized her fear was not logical or rational yet it was very real and she feared she would leave this man over her irrational fear.

When Bob would walk in the door on time she was fine. If he was even a minute late she would be near tears, angry and inconsolable, ready to fight. She did not understand nor could she express what she was really feeling. During those times, she felt like something got stuck in her throat. This was very confusing to her because as a high school teacher normally her conversations flowed with ease.

In the initial session, we worked on tools to gain immediate control of the symptoms and thus give her more clarity to get to the source of the problem. As a person begins to meditate they become more accurate and clear about what it is they are really feeling. Monique became more aware that it felt like her throat was closing down during her moments of anger.

I decided to let her use color as an indirect way to scan her body to see what emotional issues were causing her discomfort. Monique entered the hypnotic state easily; I began by having her imagine the parts of her body that were free of symptoms. I would name a body part and she would tell me the color she saw that represented that part. On her right hand a green light appeared, on the left shoulder, a blue light. At her knees was a yellow light and at her heart center, a rose-colored light. We then went to her throat, which I knew was the area where she was having the most difficulty. Monique saw her throat as half-green, half-black. Normally, in my experience, it is unusual to see an area half healthy and half very unhealthy. More often a single color shows up.

I knew something significant happened to cause that energy blockage. It also explained why sometimes Monique was at ease in her expression and other times not only was communication blocked, but also, she experienced great discomfort, emotional pain, and even rage.

We regressed Monique back to the time when her throat was a perfectly healthy color. She was 12 years old, in the sixth grade and she saw her throat energy as a white light. We moved her forward in time, frame by frame, until the color began to shift.

The white color shifted to a muted brown. I asked Monique, "Are you inside or outside? Are you alone or with someone? Is it daytime or night time? Who are you with?" I also asked her other identifying questions.

It was daytime, she was inside and her father had just come home from work. He was late and drunk again. He stumbled in the door and immediately began barking orders to the children.

He told Monique she was irresponsible because as the oldest she had not done her chores.

One would think Monique was upset because he was drunk, but actually what was really making her angry was the fact that he was trying to act like a father, criticizing her and her brothers and sisters. She was absolutely furious at him for regularly stumbling in late and acting like he was being a responsible father.

For Monique he was pretending to be someone he was not. As she shouted back at him on this particular day, declaring that *he was the irresponsible one,* she received an unexpected hard slap on the face. Monique closed her mouth, sealed her anger inside and left home at age 16.

Over the years, whenever someone did not do what they had promised, especially if they said they would return at a certain time and they were late, she would swallow her emotions. When she came to see me, she had reached the point where she could no longer stuff more emotions by energetically sealing off part of her throat.

We brought Monique's father symbolically into a chair in front of her. She told him how his behavior made her feel. She poured out all the feelings of a sad, angry, and frustrated 12-year-old. We then had her father respond. In a conversation that went back and forth for several minutes, as Monique's anger dissipated, she opened her eyes of compassion. "Hearing" the words from her father's heart brought resolution and forgiveness for her.

We then brought in Bob, her boyfriend, and stood him beside her father. She saw the vast difference in their personalities and temperaments. A wave of calm acceptance washed over her.

I asked her what color her throat was now and she said, "Bright green." Monique had previously remembered some of the drama of her growing up years but had no idea of the impact those situations had on her current life.

Questions & Reflections

1. Dr. Candace Pert uses the term "bodymind intelligence." What does she mean by this term? Can you cite illustrations of your own bodymind intelligence?

2. How does our environment influence our bodymind intelligence? Can you give personal examples?

3. Give examples of emotions, such as fear, excitement, or anticipation, that deliver physical responses in your body. How do you feel, for example, before jumping into ice cold water? How do your mouth/skin/hands, etc., feel after taking a long hike, bike ride or run? After a 12-hour or longer fast? After receiving a special reward? How do these physical responses contribute to the way you feel; e.g., overwhelmed, happy, grateful, tired, weak, etc.?

4. Have you ever "talked to your body" to receive information? Explain.

5. Have you ever had premonitions about an event or situation that actually occurred? Explain.

Exercise

Think about what it is you would like to know from your body. Do you have an old injury? A chronic pain? Lack of energy? Or, is a dis-ease plaguing you?

Begin with focused breathing as you have done before. After several minutes, when you feel your body shift into a state of calm comfort and your brain is focused and directed, open your inner eyes. Imagine those eyes are right behind your natural ones.

Focus on the body part in question. If that body part was represented by a color, what would that color be? Does that color "feel good"?

(Healthy colors are individual for each person. Usually green, blue, purple, white, even yellow, may feel good. Colors such as gray, brown and black, and sometimes red, may indicate inflammation, disease, or blocked energy.)

Imagine you are traveling back in time as you continue to watch the colored part of your energy body. As if you are running a movie projector, take yourself backward in time and stop the camera when you see that body part with a different healthy color.

Where are you? How old are you?

Ask yourself a number of questions to determine the place and time when the energy shifted. Are you inside or outside? Are you alone or with someone? With whom? Is it daytime or night time? What is going on? You can give your body part a voice and begin to converse with it.

This is often the type of subconscious work we do in therapy, but you can gain valuable information about yourself as you practice these techniques and bring up information to conscious awareness.

Can you "re-frame" or re-tell the story around this energy block in order to clear it? Is there something you need to grieve? Do you have an unresolved conflict? Is there a deceased loved one you need to speak with?

You can do some clearing work on your own; other times it may be helpful to work with a trained therapist.

5

The Science Behind Our Daily Actions

We think from our conscious, rational, logical mind. We feel and act from our deeper subconscious feeling mind.

The Conscious and Subconscious Minds

It is important to remember that thinking and acting occur in two different areas of the mind. We **think** from the **logical, conscious mind** but we **act and feel** from the **emotional subconscious mind.** The **conscious mind** communicates **verbally, through thoughts and words. The subconscious mind** communicates **nonverbally through feelings and symbols.**

We spend a lot of time cramming information into our conscious mind. We have the ability to retain much of this information and draw from it when we wish, which may make us feel very smart. However, since we don't act from this part of the mind, this feeling is an illusion, which can be confirmed by some of our actual behaviors. **Our subconscious mind is where the vast part of our intelligence lies.**

The following is an example of thinking logically or rationally and acting emotionally:

Conscious Mind: I've studied hard for this test today. I know this material so well, I'm sure I'll be able to easily answer every question, even the surprise ones.

Subconscious Mind: My stomach is quaking. My hands feel cold. (The program running in my subconscious is: "This is The Big Test, pass or fail. I always mess up on tests like these, when so much counts on them.")

You look down at the test in front of you. The words and letters are a blur. You feel like you're going to pass out. You can't think straight. Somehow you manage to answer the questions but your subconscious program tells you that you've messed up.

What if the subconscious mind were delivering a different action/emotional message? What if you walked into the exam room feeling calm and self-confident? Your head feels clear and your whole mind-body feels energized and ready to take this exam because *you feel so good about yourself and your abilities.* You feel rested, alert and excited to have yet another opportunity to do your best.

You have studied thousands of bits of information. It is carefully stored in your mind. The test is only a hundred questions, just a fraction of what you know. You feel calm and at ease with the awareness that you know far more than will be asked on the test. You look forward to giving back just a small part of that information.

This example is a "before" and "after" picture that I experience often in the healing room. When a person comes to me asking for help, often they want to perform at their peak levels and they don't know what is standing in the way. After the first session, usually subconscious programs or obstructions appear that need to be cleared. Possibly during their growing years, that person was told by a parent that they "didn't have what it takes to succeed." Or they may have just assumed they weren't as smart as their brother or sister. Other times, lack of self-esteem comes from indirect or misinterpreted messages. Feelings of failure may be deeply embedded in the subconscious and continue to influence the person's performance, *regardless of how competent they really are.*

An example such as this one illustrates the power of soul healing. The soul is our inner guide. When we tap into the subconscious and remove issues that are blocking clear communication between the mind-body and the soul, "soul-utions" or healing occurs.

This is the magic—and science—of mind-body healing. As a healer, I guide my clients through the process of reconnecting with their soul.

Often when I see someone in my office for what is referred to as "test anxiety," they have failed an important test several times. Deep fears may surface when someone is facing Medical Boards, Nursing Boards, Law Boards, and college exams. One time a man told me he was about to lose his job after 25 years as a Security Guard. The test made him so nervous, he couldn't shoot a gun straight at night. All of these individuals passed their tests when they removed their blocks to success.

Test anxiety is similar to performance anxiety that athletes and performers may feel. It's interesting that the individual has come so far in education and career, and at the very last test, the anxiety they had been able to push through before now, suddenly becomes a stumbling block. This is because every time an anxiety emotion is triggered, it reinforces the previous unresolved emotional conflicts. Over time, the reaction builds and becomes stronger.

We expand our capability and soul awareness when we move beyond our blocks and finally achieve what we are truly capable of.

I Need a License!

"Jody" was sent to me by the mentor she had been working with for a couple of years. She had finished her Master's degree in Counseling but had failed her oral boards three times. She knew she worked well with people but she was unable to pass the test to receive the necessary licensing. I was her last resort. I wished I had been her first option!

During the first session I trained Jody in the relaxation response. I sent her home with a CD to practice and retrain her brain to relax in response to stress. I knew as she began to drop deeper into a meditative state that the underlying reasons behind her fear would arise.

Jody reported she felt calmer about the test after the second week. In the state of hypnosis, we then brought up her feeling of sheer terror when faced with an oral examination. Jody regressed to age two. She remembered standing in front of her father and experiencing the same feeling of sheer terror. Her father was very "Big" and was screaming at her. Jody had wet her pants and her father was outraged. He made her stand in front of him and explain why she had wet her pants, all the while shouting loudly at her. At two years old, Jody was terrified and speechless. She didn't know why she wet her pants (I believe that is called an accident) and couldn't answer him.

In Jody's mind, I brought in the adult Jody to stand by the helpless, broken-hearted two-year-old and speak for her. She opened her throat and carried on a strong conversation with her father, defending and protecting the two-year-old.

Soon afterward, Jody passed her oral boards with ease.

Often the seed of fear and anxiety is much less traumatic than it was for Jody, but at the time when an incident occurs and that seed is planted, the feeling of fear or anxiety is very real. It merely takes a situation that brings up a similar feeling to trigger the previous unresolved emotion.

Dynamics of the Mind-Body

When our bodies begin forming in utero, we pick up messages or bits of data from our mother and our environment via our mother's thoughts and chemistry. If a mother is having strong emotions, the baby who shares her blood is also having similar feelings chemically. One important thing to remember is that *the baby does not understand why the mother is chemically reacting as she is.* The baby only feels the strong maternal response.

Dr. David Cheek (1912-1996), who practiced OB/GYN medicine for 40 years, used hypnosis with all of his patients and had remarkable results. Undoubtedly this is because hypnosis works with a patient at the subconscious level where the practitioner has direct access to the part of the mind that controls the patient's actions. If the baby, living solely in the mother's environment, is subjected to reactive emotions, e.g., anger, resentment, jealousy, fear, etc., it cannot help but experience the repercussions of such disruptive behavior. Predictably, Dr. Cheek observed a profound correlation between strongly reactive mothers and babies who felt rejected.

I had the privilege of studying with Dr. Cheek and after I had worked with many pregnant women in my own practice, I came to also believe that a mother's chemical soup is already putting bits of emotion into the unborn child. If we were that child, some of those early disruptive messages may have already posted interference signs in the clarity of our own sacred heart. Many years later, in the state of hypnosis, some of these emotions can actually be brought up and cleared.

The following is an excerpt from an article written by Dr. Cheek about the effect of the baby's in utero environment:

I was in Hamburg, Germany two years ago, and one of the psychologists in the group asked me to work with her as a demonstration. She said, "My mother and I have never gotten along, although I love her very much. She lives in Berlin. I try to avoid talking to her, but she calls me and we always get into arguments on the telephone. I'd like to know what [I can do] that might help in my relation to her because she's getting old; she needs me. I would like to be helpful, but I always find myself uncomfortable around her." So I invited her to come up for a demonstration.

We set up ideomotor—thought/muscle movement—ways of signaling unconscious information, and I simply asked her to go back to when she was just emerging out into the world. [You don't have to go through a long induction technique. That is so surprising for someone to try to conjure with: that they could remember their birth. It's the confusion technique of inducing hypnosis.]

She didn't have to be in a trance to begin with, but she went right in, to be there at the time of her birth. Her head turned to indicate the way her back was in relation to her mother. An arm came out when I asked, "Which arm is delivering first?" [This is a physiological memory that was imprinted by the adrenal hormones that were present, added to a lot because she was born in Berlin in 1943 when there were a lot of stimulating things happening—like bombs arriving.]

She said, "My mother is so happy she almost screams with pleasure to see this daughter of hers, this beautiful child."

I said, "Well, that sounds pretty good. How does the child feel?"

She shrugged her shoulders and said, "Na-ah," as though it were nothing.

To me, that meant there must have been something that had gone on earlier that had set the stage for her to reject her mother's joyful acceptance of her. So I asked her to go back to the time when her mother learned she was pregnant. [I have found that this is an important moment: women are happy, or they're scared, or they're mad. This emotion seems to make the memory lasting.] She signaled that she'd done so.

I said, "How does your mother feel?"

She said, "Scared," and then there was a pause. I was trying to think of what else to say, and in the pause, she said, "She doesn't want me." That was her interpretation of her mother's being scared.

Being scared has to do with survival. This is a right hemisphere type of impression. It's the psychic, spiritual side. It's very important for animals to know where the danger is, and to remember how they got out of danger before. Tremendously important. So, "scared" meant, "She doesn't want me."

Korzybski, the father of General Semantics… said: "The map is not the territory." The way you understand the territory is very different. The map only gives you colors, and maybe some lines of topography. What we hear and what we pick up in other ways, we filter out in accordance with our background of knowledge. Asking about the background of knowledge of a little spirit that has selected a mother is very helpful… I found out that she had selected her mother. I always ask about this. It helps so much psychiatrically if someone who hates his or her mother discovers that he chose her, or she chose her, in the first place. It lets them look at the possibility that there might be something else wrong, and they might be willing to reframe their impressions of their mother. So I asked about that.

"Is there another part of your mother that does want a little baby?"

I knew it would be "Yes," because I've found, as a gynecologist/obstetrician who has been concerned with fertility for 45 years, that women do not get pregnant unless there's a biological readiness for pregnancy. Now that's at a physiological level, way down deep. At a higher level, one which has to do with the environment, they might not be ready for pregnancy. And for years, I've enthusiastically supported the right to terminate an unwanted pregnancy, because, all during pregnancy, the little baby inside is picking up the feelings of its mother as to whether she really is accepting of what she's carrying, or rejecting it. It's very hard—it's almost impossible—to change a baby's attitude towards women if it has felt unwanted all during pregnancy.[44]

After we are born, in our early years from about 0 to age 5, we are still picking up messages from our family and environment. To make it simple, we could say some of these are positive messages and some may be negative. Some children grow up with a lot of positive messages. Other children grow up surrounded by negativity. It's interesting that even in a warm and loving home environment most kids pick up a mixed bag of messages. They may misinterpret a situation.

From the ages of about five to thirteen years old, we continue to take in messages but only hold onto those that are familiar to us. It doesn't matter if they're true or false. The acceptance factor is familiarity. If the messages feel familiar, we retain them.

An example of this might be a second-grade child who receives an award in school for reading. A child who feels like they're pretty smart may think, "Cool, I'm a good reader!" They have strong self-esteem and the message is believable.

Another child may receive a reading award in school and be a good reader, but they may think, "Yeah, right, the teacher felt sorry for me. It's the end of the year and I didn't get any other awards. I can't even get through dinner without spilling milk, I don't do anything well." That child will reject believing they deserve the award because the data they have retained in their brain is unfamiliar with the message they're receiving.

---

[44] http://www.tir.org/research_pub/cases/fetal-perception_memory.html

As adults, we respond in the same manner. If we are complimented on something we perceive we do well, we believe what that person is telling us. If we receive a sincere compliment for doing something for which we feel inept, we reject it. The recorded subconscious message of proficiency in that area is unfamiliar.

Throw Away Baby

Ms. B was referred to me by her physician for alcohol abuse. The doctor assured her I had 100% success with alcohol issues, so Ms. B had high expectations. At our first meeting when she came to me completely intoxicated and minimally functioning, I wish I could have said the same!

For the past two years, Ms. B drank every day, starting in the morning and continuing throughout the day. She had been drinking all day up until our meeting. If this weren't enough of a challenge, it was a three-hour commute from Ms. B's home to my office. The addition of the weekly six-hour drive to a therapy session seemed like it might put her over the top. Also, since she was in no condition to sit behind the wheel, she would need a driver. I thought it might be best for her to check into a residential rehab facility, but in her mind this was not an option at that time.

Ms. B reported she had been "just fine" until several years ago when her husband left her, took her home and business, and dragged her to court over fifty times in the past five years.

I found it unusual that she was not a drinker at all until that crisis occurred! I asked her about her history. She told me she was a "throw away" baby. She identified herself as "an adopted one."

During our first session when I guided her to a calmer more relaxed state, she felt worse. I knew that meant the bubbles of emotion that the alcohol had been numbing were rising to the surface.

In the second session, it was obvious to me that the husband's abandonment in "tossing her out" as if she were worthless retriggered her birth issues of being given away. She was completely unaware of this. We worked on the birth trauma.

Ironically she had been adopted into a very loving home. She adored her adoptive parents but was focused on the mother who "threw her away."

The next week, Ms. B had a delightful day with her elderly adoptive mother, baking as they used to do in former years. After her mother went home, however, Ms. B fell into deep despair and called me. It was then that I told her the parallel between being tossed out as a baby and her husband giving her the boot. By this time, she had already started to reduce her drinking. She understood exactly what I told her, and calmed down.

Improvement continued, even when intermingled with crises. Usually we needed a phone call between sessions in order to sustain our progress.

There were more crisis moments. The next crisis involved moving through the anger and betrayal from a member of her adoptive family.

As we began to work on healing the inner child, Ms. B reclaimed her beautiful, sensitive loving nature. She started to identify with her inherently kind and compassionate heart. Also, she began to view her birth mother's issues separately. Ultimately, she reconnected with her own beautiful soul. During the process, Ms. B opened her heart and soul to who she really was.

Each week Ms. B continued to grow by leaps and bounds. By week four, she said she was thinking about putting on makeup again. She brought her adoptive mother to the session and since I live near the ocean, they made it a two-day beach outing, walking on beaches she loved. On week five, Ms. B brought a sign with her that read "Because I can." It was her new theme.

Ms. B is a perfect example of how early influence from the outside world can dull our connection to our Divine nature. Mrs. B had had a highly successful career and life until the recently experienced feeling of being tossed away had triggered familiar unresolved feelings from long ago.

Hidden Desires

I received a call from a woman (Ms. D) on the East Coast who said she wanted to stop biting her nails. She said she would be visiting her sister, who lived in my area, and she wanted to make an appointment during that time. She called several months in advance so I knew the appointment was very important to her. Ms. D scheduled the session for the last day of her trip; that meant we had only one chance to do the work.

In taking her history, I noticed Ms. D was very proud of her career. She was bold and clear when expressing to me that she was very successful in a male-oriented profession. She was extremely pleased with her professional success and actually said she worked in a "man's world."

Ms. D reported her childhood was good and she loved her parents. She was especially close to her father. She also said that even when she had synthetic nails put on, she would bite and pull them off immediately. It was very important to her that she stop the nail biting now; she wasn't exactly sure why.

I began the session the way I normally would with a new client. I recorded some of what we did and then asked her subconscious mind if she had any blocks that were preventing her from growing her nails. This turned into a two-hour session.

In the state of hypnosis, I led her in a guided meditation that eventually brought her to sitting quietly in a kayak on a very smooth large lake. I told Ms. D to reach over the side of the boat with her left hand, that she would pull up an object or symbol that would give her a clue if there were any blocks to letting her nails grow with ease. She pulled up a small metal object. Initially, she had no idea what that object was. It took some questioning to identify it as a small automobile part.

In delving into the significance of that symbol, Ms. D mentioned that she used to work on cars with her father. After numerous other questions about the auto part and her father, she had a strong intuitive hit. She sat upright, opened her eyes and proclaimed, "I want a baby!"

We were both taken off guard. The awareness popped into her mind that she bit her nails to maintain a masculine image and fit into a man's world. The short nails kept her from feeling feminine, but in the deepest desire of her heart, she was in my office to give herself permission to be a woman, including having a baby!!!!

Imagine that you are obsessed with your nails, not recognizing that they represent your femininity. You walk out of a session in touch with your soul that wants to embrace your femininity and have a baby. During the session, Ms. D gave herself permission to still work in a man's world yet be a woman. This is a clear example of a deep-seated soul issue in disguise, right under her fingernails!!!

He Creeps Me Out

"Paula" presented as a beautiful woman, the mother of three grown children, who was distraught because she wanted to leave her husband of 28 years. Something about him "creeped her out," she said. Although he was a handsome man, she felt no attraction for him. He was also hard-working and financially successful. This caused Paula to feel guilty about divorcing him; she was scared about her financial future. She could not find a logical reason for her feelings, yet they continued to haunt her. Eventually she followed through on divorce proceedings, still without understanding her feelings of aversion toward her husband.

In the midst of the proceedings, information surfaced that helped her to see that something beyond logic was at work here. Her "creepy" feelings were from a source different than she ever would have imagined. As she searched her soul and stayed with her authentic feelings, she began to understand that her Personal Guidance System had been at work, warning her of something that at a logical or conscious level she wasn't aware of. Paula realized that her uncomfortable feelings had been surfacing to protect her. She listened to her inner guidance and left the relationship just in time.

What Did We Know Before?

Have you ever studied a subject or "learned" something new that you already seemed to know about? Apparently, all you needed was a gentle nudge in order to access the right "file" inside, where it was stored.

We seem to see much of this in the younger generation. Computer technology and anything related to electronics seems to come naturally to many of them. My young grandchildren are already able to troubleshoot my computer issues. No one ever taught them how to do this.

As you recognize the link between your heart and your soul, you gain access to a "knowingness" that runs very deep. This knowingness is much more significant to your soul growth than simply operating a computer.

Strong Emotions

Sometimes another person's attitude or something they do triggers a strong emotion in us. We call such an experience a "projection." Is there anyone who annoys you? Is it something they do or do not do? When a strong emotion is triggered, ask yourself if there is anything about this behavior that is a mirror for something inside of yourself. You can let this reaction show you a hidden aspect of yourself.

As author and life coach Martha Beck says, "If you spot it, you've got it"![45]

It is much easier for us to see something inside of someone else than to recognize it is something we own. The next time someone annoys you, ask yourself, "What is it about this person or their behavior that is a mirror for something inside of me?" Your enemies and those who challenge you can be your best teachers. As you learn to love them, you will increase your self-love.

'I Am Disrespected'

"April" came into my office for her second appointment. She had reported more clarity and calm in her life after her first session. Her issue pertained to the disrespect she felt on a regular basis from a family member. She had agreed to care for her young cousin for a week while the child's parents were away. The child had some medical issues and it was a bit of a commute to pick her up from elementary school, but April loved the child and was happy to help. The problem was that every time she did a favor for the parents of this child, she felt disrespected.

In this instance, she went to pick up the child from school and she arrived a few minutes later than usual. To her chagrin she discovered the child was no longer there. The parents had come home a day early and didn't call to tell her. This pattern had already repeated itself several times in the past and it was making her very frustrated and angry.

We worked on April's feelings of being disrespected. Logically, it could certainly be perceived as rude or thoughtless for the parents not to let her know they were returning home early.

I regressed April to the first time she felt disrespected. It went back to an instance in April's life as a young girl. April was surprised to draw up the memory of her father being regularly

---

[45]http://marthabeck.com/2012/01/projection-when-what-you-spot-is-what-youve-got/

inconsistent about coming to her home to pick her up for visitation at the time he said he would be arriving.

We let eight-year-old April address her father. She told him he was not honest or dependable. She told him he made excuses rather than do what he said he would. As April's eight-year-old self addressed her father in session, she was able to speak her truth, release her anger, and see him with compassionate eyes. He was inconsistent and undependable and he was also an alcoholic. She healed her heart regarding this issue with him.

April realized that part of her reaction to the cousin was unresolved anger with her father. He regularly made excuses just as her cousin had done recently. April was now better able to view the current situation regarding her cousin with a compassionate heart.

As we continued to work and I asked April questions, she considered that maybe this couple came home early because they had some personal issues. She saw the parallel between her cousin and her father. She knew they were doing their best and merely falling short, like her father. This new focus expanded her love for the entire family.

The parallel between the two circumstances: the father's undependability and making excuses, and the cousin's unpredictability, i.e., showing up without prior notification, conveyed to April identical feelings of being disrespected. Transferred from past to present, April's emotion was now directed solely at her cousin. Yet once she let go of the stored emotions with her father, she was able to easily reframe her current situation.

'I Feel Persecuted'

"Matthew" came into my office reporting that he felt persecuted and tormented. He was hypersensitive to noise and perceived sounds, such as someone typing on a keyboard, to be an attack directed at him. He said when the woman in the cubicle next to him typed, he felt extreme rage. He was worried about controlling himself. He felt bad about his rage and his feelings of being persecuted.

Extreme rage is certainly not a safe situation. In spite of Matthew's angry reactions and aggressive feelings, he never acted on them. I sensed his gentle nature and fear of hurting anyone. He'd had these types of reactions in previous jobs. When it became too uncomfortable, he would quit the job, assuming it was the situation that was causing his anger. Matthew was convinced that he needed to quit this job also.

When I asked about his past, I learned that Matthew's father died when he was four. He didn't believe this had anything to do with his current situation because as a four-year-old, he didn't remember that time.

During the first session, we worked on tools to control his extreme agitation. I sent him home with a recording to train him to quickly dissipate rage and begin to identify the emotions deeper than the apparent ones that were surfacing.

In the second session, we worked on Matthew's hypersensitive hearing. In the third session, the core issue surfaced. The source of his anger was his father's death.

Matthew was confused as a small child when his father died. He was unable to attend his father's funeral or have any type of a closure regarding his father's sudden disappearance.

In session, we brought in both Matthew's father and mother, who were deceased. He felt their tremendous love and support.

In addition, in hypnosis it was clear that part of the rage Matthew felt was related to his grandfather. When Matthew's father died, it was his grandfather who stepped up to the plate and took over as a father. Matthew cried a lot as a child and was repeatedly called a baby by his grandfather when he expressed emotion. Matthew learned that instead of appropriately releasing and working through the grief over loss of his father, he'd kept it inside. He'd held so much pain inside for so long, even the noise of someone typing at a keyboard triggered internal rage.

We then allowed four-year-old Matthew to address his well-meaning grandfather. He could now view his grandfather with a compassionate heart and realize he was trying to protect Matthew from being made fun of in school. Matthew loved his grandfather, who had become like a father to him when his own father died. It was the grandfather who would not let Matthew "act like a baby" and cry about his sadness. Matthew had completely forgotten about his grandfather's harshness because he stopped crying and as a very young boy, started to hold in his emotion.

After that session Matthew was able to feel his own gentle sensitive nature and he was filled with peace and joy.

We finished with adult Matthew talking to four-year-old Matthew. Adult Matthew told little Matthew how smart, kind, sensitive and loving he was. He also told him he was a very charming little boy! I could hear the sincerity, self-love and respect in his voice. His anger, rage and anxiety were completely released.

It was a beautiful transformation to witness. At the end of the session, Matthew was in a state of calmness and peace.

Questions & Reflections

1. If you could change one aspect of yourself for the better, what would that be?

2. Do you find yourself reacting strongly to certain situations or circumstances?

3. What are the deeper feelings in your body when you react? Could they be linked to an earlier time when you had the same feeling under a different circumstance? Explain.

Exercises

Think of a time when someone really annoyed you. Remember what was said or done and *what you felt like.* Bring up the feelings in your body.

What does your heart feel like? What does your gut feel like? Is your throat open or closed?

Do your feet want to run or are they stuck to the floor? What does your head feel like?

Bring up and recognize the strong feelings in your body.

Let yourself begin to spell out loud a word that describes the feeling.

Example:

B-E-T-R-A-Y-E-D. **Betrayed.**

Now create some sentences using that word and see what information you bring up. Keep coming up with new sentences for the feeling until you have an *Aha!* insight.

Example:

I feel (state your feeling) when...
I feel betrayed when...
I feel betrayed when...
I feel betrayed when...
I feel betrayed when...
I feel betrayed when...

One of the answers you come up with may be the key to understanding and releasing the emotion.

How Do We Examine our Belief Systems?

What are your beliefs about yourself, your potential, and your dreams? Make a list of what you believe about yourself.

As I began this chapter I did the very same exercise. Do you know what I found out? As a famous cigarette commercial used to say, "You've come a long way, baby!"

Here's my list:

- I love people.
- I am tenacious.
- I am hard working.
- I am sensitive.
- I push through and make things happen.
- I care about other people's feelings.
- I try to do the right thing.
- My role as a mother and grandmother is my most important work.
- I am committed to caring for my body.
- I love to learn.
- I see the humor in life.
- I live by faith.

These have always been my core beliefs, but at times I didn't acknowledge them. I felt like I constantly needed to push harder and do better. I allowed others who I now realized were threatened by my power, to push me down and make me feel "less than."

When I was a child I believed I was very average. I was totally bored with school and daydreamed a lot. I thought a B average made me not very smart, certainly not as smart as my friends, who got straight A's. When I went to college I surprised myself. I was not bored, I paid attention and got A's. My parents were shocked. They didn't expect that from me. As the oldest of seven children, they were grateful I was getting by without much outside assistance. **During my college years, I changed my belief about myself.**

I hope you enjoy this exercise. Look at your list. What items on your list do you want to keep because they are genuine parts of you? If you find anything on your list is "less than," where do you think you came up with that belief? Did it belong to a teacher, parent or sibling? **Do your beliefs honor you as the eternal being you are? Do your beliefs allow you to be limitless and soar?**

Strong Reactions

- Think of the last time you reacted strongly to someone, something or a situation. Remember it in detail.
- What did the emotion feel like?
- What did your heart feel like? Was it beating fast or slow?
- What did your stomach feel like? Did it feel like it was "in a knot," squeamish, nauseous, tight?
- What did your head feel like? Did you feel pressure or numbness? Did your head feel like it was spinning?
- What did your throat feel like? Was it open or closed?
- What did your feet feel like? Did you want to run or were you stuck to the floor?

Now that you have that strong emotion in your body, let your mind drift backward to an earlier time and place, to different circumstances, with the same feeling. You can drift even further back to the very first time and place when you experienced this feeling. That will be the seed of the emotion that has been fed by similar feelings and different circumstances over time.

In therapy when arriving at that place, we do whatever work is necessary to reframe the trauma. If you go back into childhood you can bring your adult self to speak for the child with wisdom and maturity.

Some of this emotional clearing is simple when you recognize you are reacting to more than the current situation. You can take in a deep breath and let go of the current reaction.

If you find yourself traveling to a place of deep-seated pain, it may be best to work with a therapist who specializes in regression work.

Visualization

Bring yourself into a quiet meditative state with controlled breathing. Calm and quiet your body. Let your mind begin to drift backward in time. Find your five-year-old self. You are getting ready to start school. What do you want that little one to know in order to feel confident?

Be the adult sensing what that child needs. Imagine you could take that child's hand and infuse them with the vision of the perfect amazing child they are. Inspire them with limitless possibility.

They are just getting ready to begin formal schooling to help them discover their interests, develop their talents and evolve their soul. Let them know they are more than enough, exactly as they are.

6

Soul Encounters – Scientific Evidence

We are souls having a human experience. We are not alone, ever.

Pulitzer Prize winning author and professor of Psychiatry at Harvard Medical School John E. Mack, M.D. points out that "we are now witnessing a complete coming together of science, psychology and spirituality after centuries of ideological and disciplinary fragmentation:

> Both modern physics and depth psychology are revealing to us a universe in which... all that we can perceive around us is connected by resonances, both physical and non-physical, that can make the possibility of universal justice, truth, and love more than a utopian fantasy.
>
> At the heart of this possibility is what in the Western secular world are called "non-ordinary" states of conscious-ness, but in the world's religious traditions is variously called primary religious feeling, mystical oneness, connection with the ground of being, or universal love... At the heart of these states of consciousness or being is a potential extension of the self beyond its usual boundaries.[46]

---

[46] https://books.google.com/books?id=EmBBp5YCev8C&pg=PA17&lpg=PA17&dq=We+are+now+witnessing+a+complete+coming+together+of+science

A Fine Line

As a nurse, the first time I held a dying patient in my arms and prayed with her, I witnessed the fine line between physical death and rebirth. This experience helped me to see and feel beyond the physical world.

Curiously, of all the nurses on the staff, I was the one who was usually asked to usher our patients "off" this earth plane. Looking back at this time—I was only 23 years old, one of the youngest and least experienced—I realize now that I was surprisingly comfortable with that privilege and task. I would even volunteer and take over for the other nurses when one of their patients was dying.

In retrospect, I wonder what made the other nurses uncomfortable or possibly frightened about witnessing someone's death. Was it because these patients were saying goodbye forever and they didn't want to experience such finality? Did this cause them to immediately identify with their own and their loved ones' demise?

Why did I feel differently about this experience? To be perfectly fair to my nurse colleagues, this was in the 1970s, at the beginning of the emergence of Dr. Elizabeth Kubler Ross's conscious dying work,[47] and also before Dr. Raymond Moody came on the scene with his authentic reports of Near-Death Experiences.[48] Death

---

,+psychology+and+spirituality&source=bl&ots=HbBz3ShoyZ&sig=te1GQZO Yu8GfMcJNN5wr_ptoZno&hl=en&sa=X&ved=0ahUKEwj07uXvpv3KAhVK22 MKHXbbA0oQ6AEIIzAB#v=onepage&q=We%20are%20now%20witnessi ng%20a%20complete%20coming%20together%20of%20science%2C%20 psychology%20and%20spirituality&f=false, http://johnemackinstitute.org/ Deeper_Causes_2003_06_IONS.pdf

[47] Kubler-Ross, Elisabeth, *On Death and Dying: What the Dying Have to each Doctors, Nurses, Clergy and Their Own Families. Scribner's reprint, 2014.* http://www.biography.com/people/elisabeth-kubler-ross-262762# pioneering-psychologist

[48] Life after Life - http://www.lifeafterlife.com/

and dying were subjects that were not usually discussed. Today we have become outrageously open and transparent about many topics that were still taboo during that era. It is perfectly understandable that the nurses with whom I worked were a bit queasy when confronted with patients who were about to take their last breath.

Also, at that time, there was no in-home hospice. When a cancer patient needed treatment or was dying, they were brought to the hospital. Some of those patients had no family or friends for support so we became their surrogate shepherds to the other side.

As nurses, we carefully titered the pain medicine, drop by drop into the IV, to keep our patients as comfortable, alert and clear as possible for their upcoming passing. Often we talked and prayed with them throughout those final moments.

During that first year as a nurse, I had encounters with Spirit that were not logical or rational. Particularly around the time of a death, I began to recognize and feel gentle guidance and spiritual protection. I knew I wasn't working alone. In those moments of intimacy with death, time seemed to stand still. These experiences are etched deeply in my mind. I will never forget looking into the eyes of some of my patients who were taking their very last breath. I recognize now that it was unusual for me to have cared for so many individuals who were completely alert and conscious until their moment of letting go. Was their Spirit passing the way it did, for the purpose of my soul growth? Or was it because I able to help them make a transition that was free of fear?

Although a myriad of other people had already recounted experiences that validated the soul's immortality and actual existence of the spiritual world, until Dr. Schwartz appeared on the scene, these narratives were considered merely subjective or anecdotal. As such, they could not be measured or quantified so they would not pass the test as scientific evidence.

I think you will appreciate the following quote from Dr. Schwartz's book, *The Sacred Promise: How Science is Discovering Spirit's Collaboration:*

> Today, along with the increased separation of church and state, the idea of connecting directly with Spirit is viewed more as a myth or superstition, or an expression of the misguided experiences of New Age flakes.[49]

Dr. Schwartz began his work at Yale. As a graduate student, he was given the assignment of studying the phenomenon of the survival of consciousness. At that time, the medical community was noticing that sometimes when a recipient received a new organ from the donor, the recipient with the new organ had thoughts, ideas or memories afterward that were clearly not their own.

In *The Sacred Promise*, Dr. Schwartz relates a story about a woman patient who was a vegan and didn't drink alcohol. This woman woke up from a kidney transplant and wanted a cheeseburger and a beer—and she could describe what they tasted like. Dr. Schwartz considered that it would surely be interesting to examine the medical history and lifestyle habits of the person from whom she obtained her kidney.

At the time, Dr. Schwartz didn't have any particular religious beliefs or commitments. He was a self-proclaimed agnostic but as a scientist he was surely open to accept whatever results his findings produced. His work on the survival of consciousness eventually evolved to become soul research.

A client of mine who had a heart transplant reported that on his way home from the hospital after the transplant, he and his wife stopped for pizza. He wanted to order the pizza with every type of meat available. He said meat was all he could think about. His wife was very upset. Previously they had always eaten

---

[49] Op. Cit., Schwartz, Gary, *The Sacred Promise.*

vegetarian pizza. She refused to let him order the meat pizza. He said eventually the craving for meat passed.

Another person who had a lung transplant told me she can tell the difference between her heartfelt feelings of grief and those of her donor. She says when she is upset with a deep heartfelt emotion, her donor, a young man, reacts. She can feel it in her chest. Her story is purely subjective, but I see consistency in these reports. I have also personally heard other remarkable stories similar to those documented by Dr. Schwartz.

Dr. Schwartz has created equipment that can record measurements as minuscule as a single photon of light, the smallest amount of energy that is capable of being measured. He uses this equipment to measure soul or spirit. The computer measures the intensity, frequency and specific patterns of the light photons.

I've discovered that anyone with an even more highly sensitive instrument, viz., our own human body and a clear mind, can also measure to a certain extent what Dr. Schwartz measures with his highly sensitive equipment.

Ryan

In 1983, I lost a son, Ryan Thomas, one of my quadruplets. The four babies were premature, so they spent their first weeks in an intensive care nursery or NICU that specializes in the care of ill or premature newborn infants. Ryan was only eight days old when he died.

One day, five years later, I was returning from a doctor's appointment in Loma Linda, California. I was barely pregnant with another child and absolutely exhausted. I stopped at a grocery store and was at the checkout counter paying for my groceries when a handsome young man, one of the baggers and stock clerks, moved my grocery bags from the counter into the cart.

I looked at this young man and at once I experienced a deep soul recognition. His face shone and I felt deep love and peace as well as an overwhelming sense of being cared for.

The young man said he would take my grocery cart to my car while I paid for the groceries, so I gave him my car keys. As he headed out the door, going straight to my vehicle, I realized he could not have known where it was in the crowded parking lot. (This was before we had remote key devices that identify our vehicles by either flashing the head and tail lights or honking the horn.)

I chased him out the door, calling, "You don't know where my car is!" Although he was headed directly for the vehicle in the crowded lot, he grinned at me and let me point it out.

I went back into the store and paid for my groceries. As I climbed into my fully loaded vehicle and the young man handed me the keys, I felt a deep sense of familiarity with him.

Before starting the car, I sat there for a few moments, realizing that something beyond reason had just occurred. I scurried back into the store, thinking I could catch the young man before he started to bag more groceries, but he was nowhere to be found. I asked all the cashiers if they could help me find him. I described him in detail, his height, age, hair color, etc. They all looked at me, puzzled, insisting that no one by that description worked there.

I talked to the clerk who was there when the young man took my cart and she also looked puzzled. After scouring the entire store, I decided to check out the receiving area at the back where the produce comes in. He wasn't there. Nowhere in the store was there a sign of the young man who had just touched my life.

Never again did I have a bagging clerk ask for my car keys and load my groceries into the car.

At the time I recognized it as a tender mercy, just what I needed in my exhausted state. I felt loved and cared for. I also felt like the young man in some way was a messenger from my lost son because thoughts of him stayed present in my mind.

I've had other occasions when I have been aware of Ryan's presence. They are always unexpected but I know exactly when they happen because each encounter fills the depth of my soul with a profound love, peace and awe.

Sara Elizabeth

In December of 2010, I lost my oldest child. During her physical life, Sara had been a great teacher and after she passed, I was surprised to discover that she became an even better one.

From the start, Sara had been exceptional. Adopted from a small farm community in Maryland, she was six months old when she came to our family. The first two months of her earthly start with young birth parents had been rough. The young couple figured out early when trying to feed her macaroni and cheese as a newborn that they were not equipped to handle a baby. Sara was then placed in foster care with an elderly couple. Four months later, she came to our family.

As a young baby, Sara would peacefully sing to herself for hours while lying in bed. Her songs seemed to put her into a trance-like state. At barely two years old, Sara was unusually verbal and talked often about God, angels and Heaven. I falsely assumed that my great parenting had something to do with that. She would sometimes tell me things I hadn't taught her.

One day when Sara was three years old and we were taking a walk, she looked up at the clouds and pointed to the sky. With determination and longing she said, "Mommy, I want to go back. I want to go back to live with God in Heaven."

I knew that Sara knew, so how could I respond as a mother? I certainly didn't want to discourage her love of God and her inner knowing. I did respond to her deep wish by saying, "Honey, we all want to go back, but God has work for us here now. You have a big job as the sister to three surviving quadruplet baby siblings." At the time, Kristina, Jon, and Matthew were only eight months old. I needed all the help I could get, even from a three-year-old.

Sara seemed okay with my response. It was then that I recognized that her verbal and spiritual gifts were hers alone, independent of me.

Growing up, Sara had many challenges. It wasn't until she was 29 that she finally received the diagnosis of Asperger's Syndrome.[50] Her years of struggle and confusion finally began to make sense to both herself and all of us.

Throughout her life, Sara spent hours meditating and talking with the angels. She would tell me about her experiences. When she was young she thought everyone saw auras and talked with angels like she did. Not until she was older did she realize her gift. Her extreme sensitivity to sound, in addition to her auditory processing disorder, was a challenge for her in a noisy household with five younger siblings.

As an adult and now on her own, Sara called me every night at 9:00. Our communication and connection were important to her.

Imagine my surprise when I discovered after she died that she still wanted to communicate with me—and not only with me, but also with others who loved her.

It came as quite a shock to me that within two days after she died, Michael, the man whom Sara adopted as her daddy and who was deeply connected to her, said to me, "Sara wants to talk

---

[50] http://www.webmd.com/brain/autism/mental-health-aspergers-syndrome?page=1#1

to you, and she wants it to be in your office." I was typing her eulogy at the time.

Sara had committed suicide shortly after she received the Asperger's diagnosis. Actually, she was the one who first diagnosed her condition, in spite of working with M.D.'s and Psychiatrists over the years. Sara believed that her condition of Asperger's would prevent her from effectively helping people. She had always longed to be of service to others.

One aspect of Asperger's is black and white thinking. Sara certainly had that. Once, when she wanted to lose weight, she quit eating for 42 days and landed in the ER with irregular heartbeats. There was no reasoning with her if her mind was set on something.

Her decision to opt out of her body so she could check into a new one and help people was misguided at best. I was shocked and devastated. After battling with depression most of her life, during her last year, she was happier than she had ever been. Sara had also promised me that she wouldn't ever kill herself.

I was completely shocked and taken off guard by the message, delivered through Sara's beloved daddy, that she wanted to talk to me. I was also very upset that she had done this to her siblings and me on Christmas Day.

I said aloud, "Sara, you're going to have to wait, I'm finishing your eulogy." I did finish it first. Then I went outside to my office, which is on the same property as my home.

It was a cold, damp, starless winter night. I lit a candle and sat in my meditation chair. In silence, I waited and waited. Nothing but silence permeated the icy cold air. I asked aloud, "Sara, are you here?"

The candle off to the side of the room flickered.

"That's strange," I thought. Maybe my breath in exhalation had traveled to that side of the room, causing the flame to flicker.

More time passed. More silence. Again I asked, "Sara, are you here?"

More silence, more cold, more waiting, another flicker of the candle.

I actually expected Sara to appear to me, so I continued to wait. Finally, a third time I said out loud, "Sara, are you here? If you are here, how am I going to know it?"

Before I could finish my thought, the candle off to the side began to sputter, hiss and flicker. I clearly picked up the feeling of her frustration.

"Oh, I get it. You're here!" I declared out loud. Once I acknowledged her presence, the crystal clear conversation began.

A part of my brain acted as an observer to the entire situation. The mother in me had much to ask and say. My daughter also had much to tell.

I remember thinking that I found it so interesting that our relationship as mother and daughter was exactly the same as before she passed. We had unfinished business. There were a few things I asked that Sara said she didn't want to answer.

I asked her why she had insisted that we talk in my office—it was so cold!

She said, "You listen better out here." That took me aback. She was certainly right about that!

Our initial conversation lasted about 45 minutes. I returned to the house, exhausted.

These office conversations continued for a few months. I began recording what we were saying on audiotapes so I could remember all of it. Sometimes she shared information directed to her siblings.

After those focused sessions, I got better at picking up her messages, which were always incredibly loving. In her earthly life when she was dealing with her Asperger's traits, there had been many challenges. At times, she was easy, sweet and tender. Other times, her emotions were unpredictable. As we communicated over those months, I sensed only her easy, sweet, tenderness, her thoughtfulness and playful sense of humor.

Sara was very close to her brother Ryan. He was only eight days old when he died, so we never had an opportunity to develop a verbal dialogue with each other. Consequently, after he left, I didn't have conversations with him as I had with Sara. But I did have experiences when I felt his love and watchful care for me. For about ten years after he died, both Sara and I would occasionally sense the presence of his soul during family times. It was so interesting that both of us would sense Ryan's presence at the same time.

We were the only ones in the family who had this awareness. Sara would say to me, "Mom, Ryan is here, right over there." Her spoken word confirmed what I had also felt. It was reassuring.

During her entire 29 years, Sara talked about Ryan regularly and claimed him as her very best friend. Even though she was less than three years old when he was born and for the eight days that he was in a physical body he lived in the NICU unit at the hospital, Sara felt a deep soul kinship with him.

Love is the absolute energy that keeps us connected, whether in or out of our body.

What Is Real?

We create our personal reality with our thoughts. Thoughts are simply thoughts unless we specifically choose them to create our reality. Our present thoughts build our future and our biology or our nature as living creatures. Where we are today is a sum of yesterday's thoughts. What do you want to create and feel in the future? What do you need to think today to do that?

If we believe we are love and are loved and lovable, the emotion of love will be what we experience in the world.

Outside Validation Is Nice but Unnecessary

It is validating and reassuring at times to be able to have some of our experiences that are unusual or that we wonder about, confirmed by a second or third party. Fortunately, several of my children have the same gift of sensitivity.

It is important to know with whom you can share these treasured moments. Have you ever had a shared personal spiritual experience confirmed by someone you know?

In my years of intuitive training, the very best lesson I learned was to **trust the information I got,** whether or not it was logical. During those years of training, often I was paired with another healer and we worked together with a single client who wanted help with an issue or some type of energy balancing. Some of this work was conducted "chair to chair," each of us seated facing the other. Other times it was done in silence, with the client on a massage table.

The beautiful part of this type of work is that when you hear your healing team member describe aloud what you are seeing, sensing, feeling or hearing, it validates your trust that you are not just making up the experience.

This gives us more confidence to tune into our vibes and *trust them.* Formal training is great, but of far greater importance is *learning to trust your own information.*

The Science of the Soul has provided copious evidence that **there is no death.** My firm conviction supported by authentic evidence of the soul's immortality is at the heart of my healing work. I have found a soulution for alleviating the fear and anxiety that most people usually feel about dying, which, as I mentioned in the Preface, is the reason for writing this book. I wanted to share this soulution with everyone.

Spirit usually communicates in quiet subtle ways; we need to pay attention and trust. Occasionally the message is direct and clear.

During a period of one of the several dark nights of my soul when I was a single mother with six children, mostly young teens, I received one of those direct and clear messages.

Looking back, I'm sure it was the grace of God that helped us make it through those times. I was outnumbered six to one. I sometimes wondered if my prayers were being heard.

Early each morning I would rise and take a long run out on a country road near my home. This would start my day in peace and gave me calm, meditative prayer time. On this particular morning run, I was feeling extremely distressed. I felt guilty and inadequate as a mother. Many days it seemed like we were just in survival mode. I knew my children deserved much more. I had concerns about several of the children struggling in school and I was absolutely exhausted, mentally, emotionally and physically. I felt very alone.

I ran fast and prayed with fervor. Hoping to feel some comfort or reassurance, I asked over and over, "Do you know I am here? Are you watching out for us?" I ran the three miles out of town and turned around, retracing my path so I would be home in time to wake up the children, give them breakfast and get them ready for school.

Suddenly I noticed a small object poking up from the dirt directly ahead on my path. I hadn't seen anything there a few minutes earlier when I previously passed that spot. As I approached the object, I saw that it was a brand new rectangular leather key chain. The leather was tipped like a road sign pointing directly at me as I approached, so I could read its embossed message.

Just a few minutes earlier, I had run past that same spot. No cars or bikes were out on the road that morning. This keychain was brand new. It didn't look like it had been tossed out of a car or had fallen off a bike. Even the metal attached to the leather was shiny and clean.

The message on the green leather was simple and direct: "Jesus Cares for You." That is all it said. What more could I possibly need? I was taken by surprise. I picked up the keychain and held it close. I still keep it on a cabinet in visible sight to remind me of the extraordinary comfort I received that day, in answer to my prayer request.

Most of the time answers are not that direct. Sometimes it is a feather, butterfly, bird, cloud, road sign, rock or license plate. Whatever it happens to be, it always comes with a subtle feeling. Pay attention to that feeling. The symbol may be something that has meaning to you.

Recently my cousin Therese told me that frogs were appearing everywhere this year to many family members. A short time ago, a young cousin of ours had died tragically. Therese told me that one day many years earlier, this young cousin and another cousin had collected numerous frogs and placed them in a container in the back of a truck.

Apparently the tub tipped over and the frogs jumped all over the little girls' legs and feet. They were horrified. To her dying day, sweet cousin April was afraid of frogs.

Shortly after her death, the family found humor and comfort in the proliferation of frogs that season; they seemed to appear out of nowhere. The family was sure the frogs were a light playful message from their dear one.

I witnessed another beautiful message a few years prior, when I was visiting those same cousins in Minnesota. That year the family lake was alive with beautiful white lilies. As we boarded a houseboat and cruised around the spring-fed lake, with my camera I snapped picture after picture of the white water lilies. Never before had I seen so many! In the past there had always been an abundance of yellow lilies and only the rare appearance of a white one.

As I marveled about the flowers, one of my older cousins, Mary, told me the story behind them. She said that over 50 years earlier when she was a young girl, she had gone with her father, my Uncle Joe, to a neighboring lake and harvested a large number of white lily bulbs. They had deposited the bulbs around the perimeter of the family lake. In those days, the farmers shared what they had with their neighbors.

During the many years that followed, only a few white lilies appeared along the border of the lake. That year was different, however. In the early fall, my Aunt Margaret and her daughter, my beloved cousin Genny, had died within a month of each other. The family felt the spectacular white lilies that appeared the following summer were a beautiful sign of love and hope from Spirit.

It's interesting that the lilies did not appear directly after the death of my Uncle Joe, who planted the lilies. Instead, they waited to bloom until after his beloved wife and daughter joined him, more than fifty years after they were planted!

This family has tremendous faith. They have also had many trials and experienced many miracles.

So, pay attention. **Ask for what you want. Believe what you see, hear and feel. The more you pay attention, the more you will witness.**

Questions & Reflections

1. Can you cite examples from events or experiences in your life that confirm that our thoughts create our reality? For example, have you ever visualized in detail something you wanted to happen, and it did in fact happen?

2. Have you had any experiences that you would call "soul encounters?" Explain and describe. Did those experiences affect your life afterward? If your answer is yes, explain.

3. How would you explain "the science of the soul" to someone who has never heard of this term before?

Exercise

If you could sit down and have a conversation with someone in your life whom you loved and who has already passed, what would you want to know?

Begin with attention to your breath. For several minutes, breathe and quiet your body. As you approach a state of calm, imagine sending a beam of love light from your heart to the person with whom you wish to connect. You might imagine it literally as a beam of loving light energy. With your inner eyes, watch the light move from your heart to the heart of your loved one. Then sit quietly and listen, feel and sense.

The information you receive will most likely be thoughts that are aligned with how that person thought.

7

Near-Death Experiences –
Scientific Validation

I was not NEAR death, I WAS DEAD!!! —Dawn, age 11

No discussion of science of the soul is complete without acknowledging the tens of thousands of reports, and perhaps even more, of Near-Death Experiences or NDEs.

Raymond Moody, M.D., is attributed with having coined the term "Near-Death Experience" after initiating serious research about the subject. In 1975, his first book, *Life after Life*, appeared, and almost overnight it became a best-seller.[51] Unquestionably, he was presenting information that was familiar to a multitude of people who may have been reluctant to disclose or even acknowledge their NDEs until Dr. Moody validated this phenomenon. In subsequent books, *The Light Beyond, Reunions, Life After Loss, Coming Back, Reflections,* and *The Last Laugh*, Dr. Moody recorded and compared the experiences of 150 persons who died, or almost died, and then recovered.

---

[51] Op. Cit. Moody, Raymond. *Life After Life: The Bestselling Original Investigation That Revealed "Near-Death Experiences."* http://www.amazon.com/Life-After-Bestselling-Investigation-Experiences/dp/006242890X/ref=tmm_pa p_swatch_0?_encoding=UTF8&qid=1455752633&sr=8-2

According to Dr. Moody, NDEs contain nine elements:

1. A Strange Sound: A buzzing, or ringing noise, while having a sense of being dead.

2. Peace and Painlessness: While people are dying, they may be in intense pain, but as soon as they leave the body, the pain vanishes and they experience peace.

3. Out-of-Body Experience: The dying often have the sensation of rising up and floating above their own body while it is surrounded by a medical team. In a state of tranquility, they observe the activities below. They experience the feeling of being in a spiritual body that appears to be a sort of living energy field.

4. The Tunnel Experience: The next experience is that of darkness and being drawn through a tunnel at an extremely high speed until reaching a realm of radiant golden-white light. Also, although they sometimes report feeling scared, they do not sense they were on the way to hell or that they "fell into it."

5. Rising Rapidly into the Heavens: Instead of a tunnel, some people report rising suddenly into the heavens and seeing the Earth and the celestial sphere as they would be seen by astronauts in space.

6. People of Light: Once on the other side of the tunnel or after they have risen into the heavens, the dying meet people who glow with an inner light. Often they find that friends and relatives who have already died are there to greet them.

7. The Being of Light: After meeting the people of light, the dying often meet a powerful spiritual being whom some have identified as God, Jesus, or some religious figure.

8. The Life Review: The Being of Light presents the dying with a panoramic review of everything they have ever done. They relive every act they have ever done to other

people, and come away feeling that love is the most important thing in life.

9. Reluctance to Return: The Being of Light sometimes tells the dying that they must return to life. Other times, they are given a choice of staying or returning. In either case, they are reluctant to return. The people who choose to return do so only because of loved ones they do not wish to leave behind.[52]

Children who have NDEs report similar experiences. Since children probably have had no exposure to information about NDEs, they would have no preconceived ideas that would influence their reports.

What would our life be like if we knew, really knew, our soul was in communication with the Divine?

Recently, I saw an 11-year-old girl in my office who DID know. "Dawn" presented in my office after radiation treatment for a brain tumor. Recently she'd had her gall bladder removed and was having great difficulty eating anything. She would try to think about a food that sounded or tasted good but she was repulsed by it when it was actually placed in front of her. Her nurse referred her to me and her mother brought her in. Dawn's father had also had a brain tumor and he had died recently. Mom was obviously very stressed.

After speaking with both mother and daughter about the issue, we dismissed mom and got to work.

I asked Dawn more questions and she surprised me with maturity that was far beyond her age. With a glowing smile she told me, "God cured me." Dawn reported that she had died when she had a severe seizure. In those precious moments, she met God.

---

[52] http://www.near-death.com/science/experts/raymond-moody.html

"Did you need to come back?" I asked her.

She nodded "yes."

I then said to her, "You had what is called a Near-Death Experience (NDE). What a gift!"

With great passion Dawn corrected me. "I was not NEAR death, I WAS DEAD!" We both laughed. "It was really hard on my mother," Dawn went on to say. Then with a twinkle in her eye, she continued, "Me and God are just like this!" She raised her hands with crossed fingers.

Then, her face glowing like an angel, Dawn told me about the angels she sees. Solemnly she said, "You know, some of them (the angels) are big and some are very small. I saw several very small angels over my mother's head when she was praying for me. My mother's guardian angel is really big. She needs that!" Again we both laughed.

This young child, wise beyond her years, knew without a doubt that she was loved and protected. She had complete trust in the God she met, and a clear knowledge of the angels present with us all. She lived in peace and joy.

It was synchronistic that on the very day we worked together, the family had a foreign exchange student move in to live with the family. The student, a vegetarian, took over the family's cooking. Dawn not only started to eat again, but she also started to help with the cooking!

Do I believe that Dawn saw God? Yes! Do I believe she is protected and cared for? Yes!

Most of us will not have the direct experience with dying and returning that Dawn did. Nevertheless, we too can know, feel the Divine guidance, and celebrate with the angels in joy just like Dawn.

Dawn's story is not unique. There are countless stories of NDEs from around the world, coming from people of every age. Dawn was young, pure, and trusting; her faith was unshakable.

Another friend of mine, Geoffrey, related his out-of-body experience during his recent quadruple bypass surgery. He said that as his heart was being worked on in surgery he felt his spirit lift up above the operating table and observe the surgeons and staff working on him. He said he saw angels assisting each staff member. He saw an angel guiding the surgeon's hands.

He said he saw three mighty angels standing together looking at him. He asked them who they were. They reported to him that they were his Personal Special Forces. They were assigned to him for this life. This had particular meaning to him as earlier in his life he had actually been in the military with Special Forces assignments. He had been shot and injured and now was aware of all of the protection that had always been there with him. He felt loved and safe. He knew his surgery would turn out just fine.

Geoffrey had an amazingly quick recovery and was home from the hospital in five days.

Physicists and scientists will testify to the fact that we are all energy beings, thus we are all connected. One factor seems to be a constant for every NDE: the energy of love is nearly always felt. I believe that even if a person has not felt unconditional love in this lifetime, when they connect with the Divine, they sense a familiarity that stirs their heart/soul connection and brings deep peace and hope.

Questions & Reflections

1. Have you ever been with a person who is close to passing and had them report to you that they were seeing other family members who had previously departed?

2. Have you ever met anyone who had an NDE? How does it affect them today?

3. How would you feel if, like Dawn, you had 100% trust that you were loved, protected and cared for? What would be different in your life today?

Exercise

Begin your focus exercise with slow, easy, purposeful breathing. Continue focusing on your breath until your body becomes quiet and calm. Imagine that your body feels lighter than air; allow yourself to drift upward above the earth. Pretend you can lift up your awareness and look down and see your body below, resting quietly and peacefully. You can see and hear everything that is going on below you. Your soul body is filled with lightness and joy. If you ever had a care or concern, it has completely and immediately vanished. You feel immersed in a beautiful light of love and indescribable peace.

A loving being of light is beside you. You sense this being knows you well and has been with you your entire life. You are greeted by your loved ones who have previously passed.

You have a familiar feeling like you are returning home; all is well. The loving light and the beings are all very familiar to you. Your previous earth life may have just felt like a moment in time. Take a moment and bask in the gentle warm light of love.

Part II – Soul Work

Scientifically Reconnecting Self & Soul

Have you ever said, "I would live my life over again if I knew what I know today?"

8

The Heart Is the Center
of Our Being

*Our day to day life can be a fine-tuning experience
similar to dialing in a very precise radio instrument
and learning what we need to do to keep aligned
with our soul path. What moment of the human life
of our soul would we want to miss?*

The Heart-Soul Connection

If there is a part of our body that knows us well, it is our heart. It
is a direct link to our soul or the spiritual aspect of our being. For
this reason, **the heart is sometimes referred to as the "seat
of the soul."** In the oral tradition and writings from our earliest
societies, including the Egyptian, Hindu and Aztec cultures, we
find reference to the belief that the soul is located in the heart.[53]

Gary Zukav's 1989 groundbreaking book, *The Seat of the Soul*,
was one of the first modern works to describe human evolution
as a path that leads from development of the personality to an
inner journey of connecting with the soul.[54]

---

[53] http://gnosticwarrior.com/the-seat-of-the-soul.html

[54] http://www.amazon.com/The-Seat-Soul-Anniversary-Edition/dp/147675
540X

In a later work, *The Heart of the Soul: Emotional Awareness,* Zukav and co-author Linda Francis explore the heart-soul connection even further. "The longest journey that you will make in your life is from your head to your heart," writes Zukav. "This is the unending journey toward spiritual growth, toward wholeness, wisdom, and compassion, and the healing that will allow you to achieve your fullest potential."[55]

Our heart is responsible for every minute of our physical life. It is our primary sensing organ, the seed of our intuition. **During the development of the body, the heart is the first organ to form.** It has been beating for us since about six weeks after conception. On the website, eurostemcell.org is the following description of the evolution of the heart:

> When an embryo is made up of only a very few cells, each cell can get the nutrients it needs directly from its surroundings. But as the cells divide and multiply to form a growing ball, it soon becomes impossible for nutrients to reach all the cells efficiently without help. The cells also produce waste that they need to get rid of. So the blood and circulatory system, powered by the heart, together form the first organ system to develop. They are essential to carry nutrients and waste around the embryo to keep its cells alive.[56]

All Issues Are Heart & Soul Issues

During my first years of practice as a therapist, as I listened to my clients' stories of the issues they wished to address, I realized if I approached these issues from the perspective of soul growth, they became meaningful, often necessary lessons for the client. Once they were understood in this perspective, often the client took a giant leap forward on their soul path.

---

[55] http://www.amazon.com/gp/product/B00109CGLW?ie=UTF8&tag=seao fthesou-20&link_code=as3&camp=211189&creative=373489&creativeASIN =B00109CGLW

[56] http://www.eurostemcell.org/factsheet/heart-our-first-organ

The issue the client presents may seem simple or straightforward, such as an addiction, overeating, anxiety, depression, fears, insecurity, or a general malaise. Likewise, our own problems may appear to be purely physical or emotional. However, as we start to delve into what is behind them—as together we breathe deeply and allow our heart to connect us to the soul—often we realize the presenting issue is simply the surface or superficial symptom, a Band-Aid covering what is beneath. The real core issue usually turns out to be entirely different.

I Want to Stop Smoking

"Jennifer," a beautiful 30-year-old woman who was a runner, told me she wished to stop smoking. She had been smoking for 10 years and she realized that the lifestyle of an athlete and a smoking habit were incompatible. Initially on the phone she assured me that her life was relatively stress-free.

Upon meeting Jennifer, she reported that her mother had died three years prior and her biggest stress was her father's upcoming wedding. She felt like he was disrespecting her mother and siblings. She could not understand how he could "get over" her mother so quickly. I sensed she was under much more stress than she realized and had unresolved grief with her mother.

Jennifer stopped smoking after one session but reported crying throughout each day and taking her anger out on her husband.

We addressed the grief and moved into the anger over her mother's death. In the state of hypnosis, we brought her mother into the room with us and began a conversation. As Jennifer addressed her mother and expressed her feelings, she could then shift her focus from anger over her mother's loss to her love and continuing relationship with her mother.

She was also able to see her father with more compassion. She realized he truly loved her mother but was uncomfortable living alone.

For Jennifer the cigarettes were merely a smokescreen preventing her from dealing with the core of her issue: grief over her mother's death and projection of that anger to her father.

Love Is a Vibrational Frequency

Like all emotions, love vibrates at a specific frequency. Writes Robert Burney, author of *Codependence: The Dance of the Wounded Souls*[57]: "Love… is our direct channel to The Source. When we can tune into that higher energy vibration we are closer to our True Selves."[58]

The Power of Love

You've probably heard it said, "Love is the most powerful force in the universe." "Love conquers all," or as John Lennon said, "Love is all you need."

According to Gary Zukav, love is the energy of the soul and there is nothing that cannot be healed by love because **there is nothing but love.**[59]

It is a fact that **love is a primal force that flows through and connects us and all of creation.** This belief is in alignment with Quantum Physics and quantum scientists' findings that **the universe consists solely of energy vibrating at different frequencies. Thus, we are energy beings radiating our unique energy signature.**[60]

As a cosmic information system that transmits love from the heart center, I propose that **God (The Divine or Source) is**

---

[57] http://www.robertburneylive.com/product/32-codependence-the-dance-of-the-wounded-souls-book

[58] http://www.abundantspirituality.com/Book-Content/higher-power-7-love-as-vibrational-frequency

[59] http://seatofthesoul.com/free-tools/authentic-power-vocabulary/

[60] http://www.collective-evolution.com/2014/09/27/this-is-the-world-of-quantum-physics-nothing-is-solid-and-everything-is-energy/

love and that love begins in our hearts. It is within our very own sacred heart that we can know the Divine, be led on our soul path, and feel the direction, synchronicities, protection and generous gifts of Love.

Louise Hay uses the word "Life" for God. She is known for her affirmation, "Life Loves You," and wrote a book bearing this title.[61]

Loving is not just a goal or ideal, it is our nature. **We are love.** Healing our wounds returns us to the truth of who we are. I remind my clients that at their very core, they are:

- Love
- Loving
- Lovable

In reconnecting with love, the universal Divine energy that fuels the universe, we tap into a powerful energy that can recharge, rebuild and heal not only our emotional lives but also our physical bodies.

Love is the pure high-octane fuel that powers us effectively and efficiently on our eternal journey. **Love expands our heart.**

Sometimes we think of learning to love as a goal, but it is much more than that. I believe it is *not so much learning to love but reconnecting with the Love that we really are,* and embracing it.

Scientific Validation for the High-Frequency Vibration of Love

In 1999, Japanese researcher Masuro Emoto began testing samples of water from various parts of the world to determine the influence of our thoughts and feelings on the molecular

---

[61] Hay, Louise, *Life Loves You: 7 Spiritual Practices to Heal Your Life* http://www. amazon.com/Life-Loves-You-Spiritual-Practices/dp/1401946143/ref= sr_1_1?ie=UTF8&qid=1455807143&sr=8-1&keywords=life+loves+you

structure of water.[62] Mr. Emoto's hypothesis was that if the human mind-body consists of 70% water, then our thoughts create our reality, both emotionally and physically.

To prove his hypothesis, Mr. Emoto froze his collected water samples and photographed them using a dark field microscope with photographic capability. Since each water crystal is unique, Mr. Emoto took 90 photos of each sample. Of the batch of 90 photos, the determining one would represent the majority. If Mr. Emoto's hypothesis was correct, clean healthy water would produce beautifully formed geometric crystals and polluted water would produce disordered particles without any pattern.

Likewise, when projecting positive, healthy, loving thoughts, words, ideas or music that is composed with the intention of producing a deep soul-connected experience (e.g., Mozart, Bach, Beethoven) on water, the resulting photographs would produce beautifully patterned crystals. When projecting negative or unhealthy, confused, hateful, destructive thoughts, words, ideas or mere cacophony (e.g., deliberately crude, offensive sounds), the photographs would show either disordered water crystal patterns or no patterns at all.

The results were astonishing to skeptics who were unschooled in the power of our thoughts, intentions and feelings and who had little information about the fact that: 1) we are energy beings whose physical, mental and emotional mind-bodies consist of a multitude of different frequency vibrations, and 2) we create our own reality, viz., the frequency vibrations of our mind-body and its environment are a direct manifestation of our thoughts, feelings, beliefs, attitudes and behaviors.

Pure clean water crystals and those infused with higher vibrational thoughts like "love" or "joy" formed beautiful

---

[62] Emoto, Masuro, *The Hidden Messages in Water.* http://www.amazon.com/Hidden-Messages-Water-Masaru-Emoto/dp/0743289803/ref=sr_1_1?ie=UTF8&qid=1455806717&sr=8-1&keywords=emoto

patterns, and those with negative thoughts like "hate" or "anger" were rough, opaque and in a state of chaos.[63]

For the rest of his life, Dr. Emoto continued to test various water samples using the same protocols. His books contain a large number of stunning reproductions of the water crystal photos when infused with spiritual energy, viz., love, joy, gratitude, health and well-being. Their sacred geometric patterns replicate throughout nature in all living forms of life and scientifically validate the high frequency vibrations of Spirit and the soul.

I remind my clients that it is just as important to think high vibrational thoughts as it is to eat high energy foods. Often we spend time on the physical self, exercising and eating organic foods, yet if our hearts are not centered in love, we will not feel optimal health or joy. In fact, deep peace and joy can be experienced even in a broken or disease-ravaged body that is filled with love.

We do not know the exact mechanism of harnessing the energy of Love, yet its healing effects cannot be denied. In healing techniques such as working in the state of hypnosis and using regression therapy as well as other forms of energetic healing, profound shifts in physical and emotional health can sometimes occur instantaneously. The power of love as a force that opens the heart and allows for free flowing communication throughout our physical-emotional-spiritual systems cannot be underestimated. Love really does make the world go around!

Love with a Capital 'L'

"Angela" was a client I had seen on and off over a period of years. A beautiful woman, she had one of the kindest hearts I have ever encountered.

---

[63] http://www.loveorabove.com/

Currently, she was homeless, living sometimes in motels or with friends. Literally she was not sure where her next meal would come from or how she would have enough money to drive to see her children, whom she dearly loved. A few days before Christmas she came to see me. During our initial conversation, she said her life was perfect but she had just a bit of discontent.

"Let's go to work and see what information this discontent is trying to give you," I said.

I then proceeded to lead Angela through a guided imagery to a large pool of water and told her she would pull out two objects that would give her information about what this discontent was about.

The first object Angela pulled up was a big gold alarm clock with two big bells. We talked about the high vibration of gold as healing energy and what an alarm clock is for—to WAKE US UP! Her discontent had appeared in order to let her know she was missing something.

I told her the next object she would pull up would be one that would tell her what she needed to do to wake up. Angela pulled up a ladder. I asked her whether she had seen a ladder such as this. Also, I asked her what a ladder was used for.

Angela said she liked to be up high so she could see. In fact, her last motel room was the highest in town.

As we progressed and discussed the significance of elevating our perspective of ourselves and our life, seeing the big picture and making choices, another beautiful symbol particularly suited to Angela surfaced. Angela realized that as she takes her loving heart to an elevated place, she can be aware of the angels in her own life. This was particularly important for her because often she felt very lonely, particularly during the Christmas season.

Angela's discontent led her to a healing that helped her remember that she was not alone, despite dismal-appearing external circumstances. Whenever she felt lonely, she could climb the ladder, elevate her awareness and remember she could always ask for help. She remembered that she was never alone.

Science of the Heart

HeartMath Institute (HMI) is one of the leading institutes conducting scientific research regarding the heart organ and its mind-body connection. Primary to HeartMath's research, in addition to studying the heart as a vital organ of the mind-body, is validation of the fact that the heart is the gateway to accessing wisdom and intelligence:

> Over the years, we have conducted many studies that have utilized many different physiological measures such as EEG (brain waves), SCL (skin conductance), ECG (heart), BP (blood pressure) and hormone levels, etc. Consistently, however, it was heart rate variability or heart rhythms that stood out as the most dynamic and reflective indicator of one's emotional states and, therefore, current stress and cognitive processes. It became clear that stressful or depleting emotions... lead to increased disorder in the higher-level brain centers and autonomic nervous system and which are reflected in the heart rhythms and adversely affects the functioning of virtually all bodily systems. This eventually led to a much deeper understanding of the neural and other communication pathways between the heart and brain.[64]

The Institute discovered that **the heart actually has its own heart brain** that "communicates with and influences the cranial brain via the nervous system, hormonal system and other pathways. These influences affect brain function and most of the body's major organs and play an important role in mental and emotional experience and the quality of our lives."[65]

---

[64] https://www.heartmath.org/research/science-of-the-heart/
[65] Ibid.

HMI's studies show that the heart radiates an electrical current 40-60 times more powerful than the brain's electrical emissions, and the heart is magnetically over 1000 times more powerful than the brain. Their research shows that when we feel genuine (heartfelt) intentions of desire, electromagnetic patterns are formed in our hearts.

Their hypothesis is that conscious creation influences templates in our molecular structure that amplify the new pattern and then harmonize and mirror the earth's magnetic frequencies, allowing us to manifest our intended desire.

Since 1991, HMI has been studying how an individual's heart-generated electromagnetic currents not only affect the individual but also people around them, and may even affect our planet. HMI now has measurable evidence that the electromagnetic field generated by the heart extends far beyond our skin.

HeartMath's Global Coherence Initiative (GCI) explores the interconnectivity of humanity with Earth's magnetic fields:

> Scientists know Earth's resonant frequencies approximate those of the brain, heart and autonomic nervous system, and studies show surprising relationships between health and behavior and solar and geomagnetic activity. Findings support GCI's hypothesis that Earth's magnetic field carries important biological information linking living systems.
>
> GCI is helping people realize the interrelation of these forces and the ever-deepening resonance and nurturing of spirit precipitated by coherent heart-based human connections. GCI seeks to demonstrate that increasing heart connections will lead to **intuitive solutions for global challenges and transformation of our world and consciousness. [Emphasis added.]** [66]

---

[66] https://www.heartmath.org/gci/

Making the Heart Connection

How can we learn to make a better heart connection?

First and foremost, **it is most important to connect with our very own heart.** The experience of love and all other heartfelt emotions really does begin with each of us. It is of primary importance to have clear access to the feelings of our own heart. **If we know our own heart, then we will better know those with whom we resonate.**

One thing you know about your heart that is obvious is that it is beating, because you are alive, in a physical body, reading this. You may wish to think of your mind-body awareness as a computer or information system that connects all of your parts. Every body part can send and receive information through your heart. It is like a Grand Central Station, an information/ transportation hub where all information meets in order to be dispersed to various destinations.

If you imagine your own heart as your center point with the ability to run all information through it, you can use it more effectively for accessing and sending information.

Physical, Mental & Emotional Benefits of a Heart Connection

- Improved health
- Increased immunity
- Peace of mind
- More energy

A Harvard University newsletter cites findings regarding the healthy benefits of strong relationships:

> Social connections... not only give us pleasure, they also influence our long-term health in ways *every bit as powerful as adequate sleep, a good diet, and not smoking.* Dozens of studies have shown that people who have satisfying relationships

with family, friends, and their community are happier, have fewer health problems, and live longer.

Conversely, a relative lack of social ties is associated with depression and later-life cognitive decline, as well as with increased mortality. One study, which examined data from more than 309,000 people, found that lack of strong relationships increased the risk of premature death from all causes by 50%—an effect on mortality risk roughly comparable to smoking up to 15 cigarettes a day, and greater than obesity and physical inactivity.[67]

Scientific research regarding the biological and chemical factors that trigger mental health issues also could influence heart disease. "The head-heart connection should be on everyone's radar," said Barry Jacobs, Psy.D., a clinical psychologist and director of Behavioral Sciences at the Crozer-Keystone Family Medicine Residency Program in Springfield, Pennsylvania. "It's not just being unhappy. It's having biochemical changes that predispose people to have other health problems, including heart problems."[68]

John Hopkins Medical School reports that certain forms of exercise or body movement such as yoga are considered extremely valuable for making a strong heart connection.[69]

"What's good for your heart is good for your brain, and what's good for your brain is good for your heart," says Nicola Finley, M.D., a physician at Canyon Ranch in Tucson.[70]

---

[67] http://www.health.harvard.edu/newsletter_article/the-health-benefits-of-strong-relationships

[68] http://www.heart.org/HEARTORG/HealthyLiving/StressManagement/HowDoesStressAffectYou/Mental-Health-and-Heart-Health_UCM_438853_Article.jsp#.VsH0uU0UW74

[69] http://www.hopkinsmedicine.org/health/healthy_heart/move_more/the-yoga-heart-connection

[70] http://www.canyonranch.com/your-health/health-healing/staying-healthy/brain-fitness/the-heart/brain-connection

Your heart pumps about 20 percent of your body's blood to your brain, nourishing it with the oxygen it needs. However, [certain behaviors and conditions] can interfere with blood flow by causing narrowing of the blood vessels and hardening of the arteries. This, of course, can cause heart problems. But a loss of blood flow can also lead to problems with thinking and memory and an overall decline in cognitive function.

...if you have heart disease, you also have an elevated risk of developing a form of cognitive impairment [a condition that impacts thinking, language and judgment] that can lead to Alzheimer's disease. There's even a link between eating a diet high in heart-clogging saturated fat and poor performance on tests of thinking and memory, according to researchers at Brigham and Women's Hospital. Another study found that a risk assessment tool for heart disease is better at predicting memory loss than a dementia risk assessment, confirming a link between heart-healthy numbers [cholesterol and blood pressure levels] and a sharp mind.

All this means that the steps you take to improve your cardiovascular health are truly doing double-duty, helping protect your brain as well as your heart.[71]

We can learn from past mistakes and consciously choose new ways of navigating these great energy influxes to create periods of human flourishing and humanitarian advances. When outdated structures that do not serve humanity collapse, an opportunity opens for them to be replaced with more suitable and sustainable models. Such positive change can affect political, economic, medical and educational systems as well as people's relationships in their workplaces, homes and communities.

Benefits of Living Heart-Centered

- Increases our capacity to more fully love and forgive.
- Increases our capacity to acknowledge, understand and express emotions.

---

[71] Ibid.

- Upgrades our ego or self-identity.
- Enables us to become more aware of our Divine nature.
- Enables us to access and download higher guidance.
- Enables us to access universal wisdom through the superconscious.
- Upgrades and heightens our intuition.
- Enables us to see the truth beyond the walls of deception.
- Enables us to have better access to our authentic self.
- Enables us to access and understand our soul path.

Our hearts are powerful energy generators and receptors. If we compare our heart energy with the energy transmitted from the brain, we learn that it is approximately 60 times stronger.[72]

Many factors could impede the heart from experiencing an easy flow of information. Blocks can take the form of pain, trauma, childhood misinterpretations; emotional energy stored as anger, victimhood, and guilt; and messages that don't resonate with our heart. If you think of emotional blocks like heavy knots in a telephone line, small knots slow things down and many knots can block a pathway of communication.

I Am Angry

"Peter," age 37, presented in my office stating that he had an anger problem. Peter's third wife was leaving him and he knew his employees hated him. The decision of the third wife to leave was Peter's motivation to seek help.

Peter reported he had always been angry. He told stories of the trouble he was to his parents as well as other people in his youth.

[72] "The heart generates the largest electromagnetic field in the body. The electrical field as measured in an electrocardiogram (ECG) is about 60 times greater in amplitude than the brain waves recorded in an electroencephalogram (EEG)." —Rollin McCraty, HeartMath Institute Director of Research, "The Energetic Heart: Bioelectromagnetic Communication Within and Between People." https://www.heartmath.org/articles-of-the-heart/science-of-the-heart/the-energetic-heart-is-unfolding/

His anger was explosive and it was getting expensive for him to have his wives divorce him.

With life-long anger you might expect therapy and healing to take many months or years, but in Peter's case it took only two sessions. During the first session we worked on tools to release anger and increase his awareness to the feelings behind the anger.

In the second session I had him bring up the feeling he had the last time he got mad. I then regressed him to the very first time he had felt that rage. I asked him a series of questions to help him identify that first episode of rage: "Are you inside or out? Is it daytime or night time? Are you alone or with someone? Give me a report."

Peter's response was, "I am behind bars like a jail. The walls are white; I am looking at a closed door." He was visibly upset.

"What is behind the door?" I asked.

"My mother," said Peter.

As the story unfolded, Peter realized the bars he saw were his crib. He had been screaming, which is not unusual for a two-year-old. Peter remembered his mother had just had a baby. Tired and frustrated, she had placed him in his crib and shut the door to his room. **He felt rejected and not listened to.**

We reframed that scenario and had mother come into two-year-old Peter's room and explain to Peter why he needed to go to his room to nap. She then reassured him that he was loved and precious to her.

Peter's transformation was remarkable. He actually looked physically different after that session. Although only in his thirties, Peter had deep lines in his face from holding so much anger. The lines were softened and Peter looked very peaceful.

He was overwhelmed with the peacefulness he felt. He kept repeating that he had never felt so good in his life, ever before.

The next week Peter returned, still with a smile, and said he had a new problem. "No one trusts me," he said. "I don't understand why!" He said his employees were actually taking bets on which one of them he was going to fire next!

I reminded Peter that he had been untrustworthy for decades and it might take a while for others to change their mindset.

Peter's lifelong issue with rage was extinguished. He had been acting just like a two-year-old behind bars every time he felt slighted or not listened to. It was a profound transformation to witness and was life-changing for Peter, his wife, children and employees. Peter discovered he had a kind and gentle heart.

Love Transcends This Earthly Plane

During those first nights when I spoke with Sara in spirit, she told me that not only is love the most powerful force in the universe but also it transcends this earthly plane.

In the first night of conversing with Sara, the observer part of me was marveling at the process I was experiencing. I told Sara that I thought this entire experience was really weird. I asked her, "How is this crystal clear communication possible? How can I communicate with you as clearly as when you were alive in your body?"

Sara showed me a picture in my mind, similar to the cover of this book. It looked like a photo of the night sky that was taken with an open lens, catching the stars in motion. Instead of pinpoints or dots, they appeared as thin strands of blue light interconnecting with each other. The threads of blue light were very crisp and three-dimensional. Sara said the lights were the energy of love that connected souls. She said, "That is how we are communicating." It was through the energy of love, our love connection.

She then went on to say, "Those of you who are on earth in your bodies can brighten and strengthen the connection with your loved ones by increasing the love you have for those who have passed over."

In other words, the strengthening of communication must start with those of us who are still in our physical bodies. If we do not feel connected, it doesn't mean those who have passed don't love us, but rather, it is *our reaching out and connecting with them* that somehow enables this connection to continue to grow.

Love truly is the most powerful, lasting force in the universe.

Children are often very open and soul-connected or sensitive to spirit. When I was very young I had three friends who were very real to me. My parents called them my "imaginary friends"; no one else could see them. I still remember what they looked like. They were small like me and playful. At times, I would refuse to let someone sit in a certain chair because one of my "friends" was seated there.

I would refer to my friends by name. This clearly amused my parents. As I grew older and heard the reports of my imaginary friends, I remembered what I saw but dismissed it because others seemed to find so much humor in my active imagination.

One day when I was about eight years old sitting quietly in church, which was probably the only quiet time in my life in those days, I discovered that if I squinted and looked at certain people, I could see light around them. I found this amusing. When I asked my mother about it, she said kindly, "Honey, it is a trick your eyes are playing on you."

This didn't stop me from sitting quietly and viewing people's auras. What a cool trick! Since no one paid any heed to my little game and it didn't seem to have any value, eventually I also dismissed it.

Sara also saw auras and could describe them in detail. I told her it was a gift and that not everyone could see them.

It isn't necessary to see auras or angels to feel connected to spirit and to be moving on your own soul path. **What *is* necessary is connecting with your heart and feeling love within, drawing you forward on your journey.**

As I expanded my clinical practice with a focus on the concept of soulutions that were sourced in love and accessed through a heart connection, I saw my clients grow exponentially faster, achieve greater inner peace and manifest at a highly accelerated rate. The business professionals I worked with saw an expanded "big picture" and deeper purpose in their lives.

If you are not receiving the love you desire, it seems like a pretty good idea to explore what is creating this state. Obviously, most of us want to place the blame for lack of love on something external to ourselves. That is a waste of time and energy, but often it feels good because blame seems to alleviate the pain, even if only briefly.

Blame energy only helps you remain out of balance, whether you're blaming yourself or someone else. "Being in balance" is based on the premise that you receive in life what you are aligned with.

I often remind others that they will create whatever they think about. A soul living in and through love cannot help but further evolve and also be a force that shifts the consciousness of other individuals as well as the planet. We are all connected. You will feel love as you give love.

If this is true, it changes everything. If it is not true and you live it as if it is in fact true, it still changes everything for the better.

Why not give it a try? The more you play with this idea, the more natural it will feel.

Questions & Reflections

1. Where are you now in relationship with yourself? Are you happy and at peace with who you are?

2. Do you see yourself beyond just your job or the roles you play?

3. How do you use the roles in your life to evolve?

4. Where are you now in your relationship with others in terms of your heart?

5. How do you feel, i.e., really feel, in your heart?

6. How did you arrive where you're at?

7. How do you want to feel—about yourself, your life, relationships?

8. What do you see when you look deep inside of your heart?

9. What do you want to see and feel in your heart?

10. If you were to think about your biggest life's lesson thus far for your soul, what would it be?

Visualization

Take a few minutes to silence your body so you can better connect with your deeper self or even the superconscious, that part of the mind that "transcends human consciousness."[73]

---

[73] The superconscious mind is "of, relating to, or possessing the highest consciousness or a margin of consciousness above that within the ordinary range of attention." It is also the place where "psychic activity" occurs. If we pursue the definition of "psychic," we learn that it is "of or relating to the **human soul** or mind; mental (opposed to physical). It pertains to or notes

For several minutes, take slow deep breaths and focus only on what it feels like for your body to breathe in, and then on exhalation, what it feels like to breathe out. Become present within your body.

Draw your awareness inward, into your lungs. Feel the comfortable oxygen-rich air move in your nose, down your throat, into your lungs and then to your heart. Fill your heart with a sense of stillness, maybe imagining the steady beat of your heart supporting you, moment by moment.

Direct your attention to your heart. Recognize as you fill your heart with oxygen that you are now supplying it with its basic need. Then imagine as you exhale, your heart pushing out the oxygen-rich blood to take care of the rest of your body.

As you continue to focus on your heart, become aware of it as the center of your being.

You might even imagine following your heartbeat back in time, all the way back to when you were a tiny baby in utero before birth. Imagine floating, warm and comfortable, with your tiny heart beating in your body's center. Imagine what that heart might feel like, pure and fresh, ready to begin the learning adventure of life, feeling safe and present in the moment.

Continue to take slow, deep, easy breaths, focusing on the air moving in and out of your lungs. Direct all awareness inward to your lungs. Follow this breathing pattern for several minutes.

---

mental phenomena." Also, psychic refers to anything "outside of natural or scientific knowledge; spiritual." (Note how our modern belief in separation between science and spirituality is even written into our dictionaries!)
Psychic further refers to "some apparently nonphysical force or agency: psychic research; psychic phenomena. A psychic is "someone who is sensitive to influences or forces of a nonphysical or supernatural nature. http://www.merriam-webster.com/dictionary/superconscious, http://dictionary.reference.com/browse/psychic

As you follow the air in and out of your lungs, remember this breath is a mirror for a pattern you are claiming in your life, that you automatically bring in to yourself all that you need— opportunities, people, circumstances—all things that serve you. You effortlessly let go of all else, as easily as you release your breath.

You may then recognize that your body knows it is essential to care for your heart first. Your lungs send the oxygen-rich air you are breathing directly to your heart.

You honor your heart and see it as your center. Become aware of your heart as a center of energy. You might imagine it as a glowing light. With each breath, you illuminate your heart, brightening it as if the oxygen you are breathing were responding the way a flame does to oxygen.

With each easy breath, your heart light becomes brighter. Become aware of the emotion of love in that light. Feel love for your miraculous heart that supports you each moment of the day. Watch that light grow and send rays of light outward from your heart to the area surrounding your body.

Send a bright, strong loving ray of light to someone close to you, alive or deceased. Now quiet your body even further and feel that soul's light returning back to you. Look for the connection of love. You may see it as light or feel it as warmth.

Expand your awareness upward as if you were miles above the earth, looking down through the nighttime sky. See the glowing dots of light from each of the beings on earth. The earth is spotted with countless heart lights. Look carefully and you will see the multitude of connections those lights have with others. You may see something that looks like the World Wide Web: all points of light reaching out and interconnecting. You are part of those lights.

Feel the strength and beauty of these connections. Notice that you are aware of the connectedness of all humans: the significance of each soul, the strength of loving light brightening the entire earth.

Bring yourself down into that intricate web of loving connection. Recognize your own light contributing to the brightness of the earth.

Note: Using this same image of heart light, you can send your loved ones a beam of your own loving energy anytime you wish. That energy of love is eternal.

You may want to flag this page of the book and return to this meditation often.

9

Allow the Divine to
Guide Your Path

*We came into this body from Source. WE ARE
Source energy. With the energy of our thoughts, the
creation process began.*

We are an extension of Source or God. Divine energy flows through
us. We think of God, Source or the Divine as non-physical. This
non-physical Source experiences constant expansion as we put
our hopes and dreams "out there." This is a very mind expanding
concept.

Before the World Wide Web

In the Fifties and Sixties when I grew up in Burbank, California,
life was relatively simple. My parents, who were part of the
World War II generation, were conservative, middle class and
hard working. Religion was built into our daily life and church
attendance was non-negotiable. We were taught to believe in
God, pray, obey the rules, and expect miracles. No one I knew
meditated, practiced yoga, or for that matter even exercised.

We ate Weber's white bread and bologna for lunch and there
was no organic labeling. We trusted that if we prayed, God

would listen. "He" was looking after us and could surely supply the missing nutrients from a bologna and white bread sandwich.

Already during our growing years, we had our future plans in place. After graduating from high school, we would go to college and then get married and have a family. In Los Angeles even those who didn't pursue higher education had great employment opportunities in the film studios or technical trades. We worked, saved money, bought a house, raised a family and retired. That was the plan, neat and simple, and when life was over, we died.

At that time, it never would have occurred to us to question life more deeply, to consider the notion of a "soul path," for example, or soul learning. We knew where we were going, didn't we? Our life in the here and now was neatly mapped out for us. The formula that our parents taught us was guaranteed to deliver the sum total of all that was good, and because it worked for them it would work for us—right?

As for soul issues, God took care of those matters. Even if we knew S/He was there to listen, what we didn't know was that *we were an integral part of that God conversation.* It would never have occurred to us to allow the Divine within us to guide our path because we hadn't consciously made that connection yet. Our minds were not primed with that awareness.

I think you can see the difference between being open and sensitive to Divine guidance—playing an active role in planning and navigating your soul path—and passively allowing parents and others to map out your life for you. Since each of us is unique, our parents' path could never be the same as ours. Our soul or spiritual essence wasn't invited to participate in the planning session, so the resulting blueprint surely wasn't our "soul path."

Welcome to the Information Age

One of the major global changes that occurred over the past several decades was the arrival of the Internet. Access to the World Wide Web through computers, smart phones and many other electronic devices expanded our horizons beyond our wildest imagination and ushered in what is known as the Information Age. The entire global community became our local neighborhood and the world was now our school, concert hall, laboratory and playground. Diversity enriched our lives both externally and internally. In the process of expanding our cultural parameters, we also started to learn more about ourselves and who we really are.

Today nothing is as cut and dried as it used to be. A college degree or vocational certification will not necessarily secure a position in our field of study. Even if we do get a job, it provides no guarantee of permanency.

In our parents' world, if a job turned out to be boring or highly stressful, quitting might not have been an option, but now we've started to look at our lives differently. Is this really how we're supposed to spend the rest of our days—doing something we don't like, or coping with a work environment that is joyless and stressful?

At some point in adulthood, we may also have admitted to ourselves that we don't really resonate with our parents' values. We may have discovered we have other talents and abilities that we prefer to develop.

Other changes also start to take place. As we discover more about ourselves and the larger world, we ask questions that never would have occurred to our parents—or if they had, possibly they would have dismissed them as frivolous or unimportant.

Possibly our parents don't approve of our lifestyle—but we're happy, aren't we? Or maybe our marriage didn't work out and we've decided to get a divorce—something some of our parents may never have condoned. With our new thinking and a different set of values, we ask ourselves: Is it better to "cope" for the rest of our life in a relationship that is joyless, or file for a divorce, allowing both parties to be free to create more exciting, productive lives?

I Look for Faults

"Bobby," a successful business executive, came into my office wondering what was wrong with him. He was in a beautiful and loving relationship. Currently he and his partner were commuting back and forth to be with each other, and they were planning to get married.

Bobby reported that when he was with his partner, things were wonderful. He was completely at peace and had never felt so loved or loving. The minute he drove away, however, doubt would creep in and his critical mind would surface. Bobby would look for cracks and flaws in his partner and go into fear mode. He lost faith in himself and his decisions.

Bobby felt terrible about these traits and was sure something was wrong with him. He wondered if he would be better off living alone and never having a relationship.

In the state of hypnosis when we brought up those critical fear-based feelings and regressed them, Bobby was surprised to go back into childhood and hear his parents "discussing" how they detested a particular group of people, some of whom frequented their business.

He was amazed to remember that his mother, in particular, could be very nice to the customers when talking with them, but as soon as they left, immediately she became critical and made extremely disparaging remarks about them.

Bobby recognized that his mother's two-faced behavior was the reason he didn't trust his partner's kindness and generosity. Although his partner had never shown any prejudice for anyone, it was when she was not present that the distrust arose.

This was a long session. Bobby addressed his mother, father, and partner. He reclaimed his true nature of looking at others and loving them through his heart. He also embraced the idea that he would accept all others different from himself, breaking the cycle of prejudice and judgment he had been taught.

Submission vs. Surrender

The definition of submission is "to yield oneself to the authority or will of another." Submission is a passive act of obesience.

The definition of "to surrender," on the other hand, is not to give up one's power but to resonate with our Divine Self—to tune into our divinity and allow ourselves to listen to the guidance, promptings, and awareness that arise from within. Our truest thoughts are always, "Thy will be done," since it is our Divine Self that is offering us guidance. If we ask questions about our life's purpose and moment by moment move ahead expecting guidance, we have it. We want the Divine within to hold the Ego's hand and walk us forward.

Enduring a "dark night of the soul" or an experience that causes us to ask, "Why is this happening to me?" requires faith, prayer and if possible, spiritual guidance. Keep a journal. Trust that all dark nights have a light at the end of the tunnel. Look for that light.

In spite of the outward symptoms that may present for us— restlessness, fear, anger, anxiety, depression—they are all simply the outward manifestation of something deeper inside that is out of balance. As we move beyond the distractions we can move toward love, compassion, empathy, forgiveness, honesty, patience, service, responsibility, relationships, and trust. By living fully present in the moment and maintaining an expanded perspective, we move through the river of life with grace and ease.

In my work often I use the symbol of water as a metaphor. We are born from water and our bodies are mostly composed of water. Comparing our soul life to a trip downstream on an easily flowing endless river can make the experience seem more effortless. At any given moment, of most importance in going downstream is to be present and to enjoy the journey. We might compare the river to our destiny. If we don't listen to our heart and if we resist the easy flow of the water, it is like swimming upstream. Life will be a constant struggle that will make us feel stuck, tired or frustrated.

The patience and trust that we are on a soul journey and that all is well can keep the journey enjoyable and manageable. We can put everything into perspective because it is only the life learning, the love, and relationships that really matter.

How Can We Trust that All Will Be Well?

None of us really has any control over the future. We think we do. We measure and carefully figure out our bank accounts to be sure we will have enough money for retirement, we fix the leaky roof and we even take vitamins. We do what we think we need to do to maintain control and safety. The reality is, none of us knows exactly what the future will hold.

Trust is both an emotional feeling and a logical belief. To trust in yourself is of primary importance. Meditation and addressing emotional issues can help us reconnect with our true selves to establish trust if it is lacking. Trust in a higher intelligence or Power is faith that is now backed by science.

Our trust in the "learning process" and the adventure of life brings peace. We trust that we will have all we need, we trust that we are not on our own in this life, and we trust that we will be in the right place at the right time. We trust that life is a learning adventure and we are learning.

Trust in your intuitive wisdom. Trust and pay attention. Trust that you are a spiritual being having a human journey—and trust that all will be well.

Trusting in your soul self and in the bigger eternal picture can lead you to purpose.

Our Work: Is It Just What We Do?

How can we use our job to further the growth of our soul? In this instance, we are defining "job" as "a paid position of regular employment," or "a task or piece of work, especially one that is paid." In other words, the work that you are doing may not be in complete alignment with your larger goals, but currently it is a way to pay your bills. It is a stepping stone. For example, a college student may take a job as a waiter or taxi driver to help pay some of their school expenses.

You can bring meaning to any job. First and foremost, regardless of the situation, *do your best.* Make a difference by connecting with your fellow workers. Enjoy getting to know them. You may discover they have much more in common with you than you originally thought. This could lead to some great conversations as well as personal growth lessons.

Probably you've heard the favorite "State Park" rule about leaving the campground in the same pristine condition in which you found it. I'd like to modify this by stating that when leaving a job, leave it *better than when you arrived* and you will discover unexpected rewards.

I Want to Be There, Not Here!

"Paul" was a successful business entrepreneur who longed to spend more time studying the healing arts. The company he had developed was providing a great service to the planet. He was very successful and he made plenty of money. Yet even though income was an important aspect of his work, he was distressed that his time was being spent just making money. He became more resentful as life went on when he wasn't able to devote time to what he really longed to do. This is a common scenario for many people who feel stuck in a job until retirement.

In the state of hypnosis, we took a look at Paul's life path and he could definitely see the healing arts in his future. The picture Paul's mind gave him when we asked for direction was one of a patch of blue sky (his freedom) beyond the clouds (his job) up ahead. The blue sky represented his goal of total freedom to do what he wanted.

As Paul brought this scene into his mind, he realized the blue sky was just a short distance (a couple of years) ahead.

Paul shifted his perspective to the blue sky (his freedom) instead of the clouds (his job). He began to see that his day to day business was really taking him to where his soul truly longed to be.

I told him he may find it interesting to discover himself loving and appreciating the challenge and creativity of running his business because he now perceived it as a pathway to his deeper goals.

We can use all circumstances, especially difficult ones, to encourage us forward on our path.

Any Job Can Be Soul Work

If you love your career or job—if it is true "soul work," i.e., in alignment with your Divine purpose—you will surely reap more rewards than you could ever imagine.

James, my fifth child, was a particular delight; I got to be with him alone at home while the older children were at school. He was a deep thinker, even at three years old. Every Tuesday morning, James loved to run out and greet the man who picked up the garbage.

As soon as he heard the truck coming down the street, James would squeal with excitement and run toward the outside door. At our house, the driver would get out of the truck and give James a warm hello and a high-five. It made James feel very special. James learned to have great respect for this man. In fact, sometimes this man would bring James toys he had picked up along the route. James treasured those toys.

One time James's friend urged me to retrieve a large helicopter sitting in the driveway a few blocks away that apparently the homeowners were throwing out. The helicopter became one of James's most prized possessions.

One might not think the job of picking up garbage is glamorous or that it could be soul expanding, but for this man and my son it surely was. Several years later, my son played football and his childhood friend, the man who picked up our garbage, was one of the coaches! James learned to look into the hearts of everyone and never to judge anyone based on their job or circumstance. He had deep respect for our garbage collector and his coach.

While working at your job—while doing anything in life—listen, ask and trust that you will be at the right place at the right time. Pay attention. By virtue of being human, you can make any job be your soul work.

Crossing with Love

One morning I dropped off my grandson, a fourth-grader, for school. I waited as he crossed the street. He had told me stories about his crossing guard, Mrs. P, and what a good friend she was to him and the other children. As Jacob crossed the street, I witnessed the crossing guard greet each child, one at a time. When a child crossed the street, Mrs. P would drop to their level, look them in the eye and give them a sincere greeting, talking to each of them as if they were the most important person in the world. It brought tears to my eyes to witness her love and kindness. I realized that for some of those children she might be the brightest light they would encounter all day.

At that moment, the job of a crossing guard seemed like the most important one in the world.

What is your work? What is your passion? How do you make a difference with your heart and your time? Are your employees, coworkers, customers and associates in a better place because of your presence?

Recognizing Your Soul Path

Some young people, especially those who have already made enough money by the time they're in their thirties to support themselves for the rest of their days, start to consider that life may have a purpose beyond mere survival. Maybe there's a bigger meaning behind it all. Increasing numbers of Millennials, our newest adults, appear to be basing their thoughts and actions on *principles* instead of *goals.* They may be searching for a way to make a difference in the world, placing purpose before money issues and survival.

Many of us who are older may reach midlife before we begin to re-evaluate our priorities. At some point we become aware that we are not living up to our full potential, or we maybe we just feel stuck. Others in midlife are so busy "just living and doing," they give no thought to the bigger picture. Sometimes it is not until a person becomes elderly before the soul is able to learn its greatest lessons. Illness or some other crisis may cause them to stop and reflect on what life is really all about.

In order to align to our authentic soul path:

1. **We recognize that something isn't working for us**; we are unhappy, feel unfulfilled, or feel like we could be more at peace. We want purpose and ease in moving through our life.

2. **We make a decision to do something about it.**

Regardless of when our soul learning starts to happen—although the earlier we begin to pay attention and acknowledge it, the better—the fact that more people are hungering for this inner connection is a sign that we are witnessing a major global transformation. Today more young people are passionate about maintaining optimal health, legislating for clean air and water and providing basic necessities for everyone living on the planet. Once we realize that we are all part of the same immortal family, our values change.

Also, when we awaken to the fact that soul learning is what we take with us when we leave our physical body—that this learning is the only part of us that is immortal—we start to view life from an expanded perspective. **It is no longer the number of days in our life but** *the quality of life in each of those days* **that is important.** Gratitude also enters into the quotient.

In my years of working with dying patients I found that even up until the last stage of earth life, some individuals had never prayed, meditated or even thought much about their soul. In those final days of letting go of their bodies, however, even the most hard-headed individuals opened their minds to listen and trust, if this experience was available for them.

In the 1970s, I worked in an Adventist hospital. The staff was very progressive and encouraged us to pray and be fully present with our dying patients. In their last moments of life, all of my patients, many of whom had no family, were open to any and all soul information I could offer. I saw many facial expressions shift from turmoil and fear to peace and acceptance. What a privilege it was to work with souls at such a sacred time of life!

The Yellow Linen Suit

At a Northern Virginia hospital where I worked, "Mason" had developed quite a reputation in the Intensive Care Unit among the staff. He was downright ornery! I'm sure the nurses who knew him 35 years ago would still remember him.

A cute little man with a black patch over one eye, whenever Mason's wife, the only love in his life, was admitted to the ICU for heart problems, he would come to visit her dressed in the bright yellow linen suit that she absolutely hated. She would get agitated with him, which wasn't good for her progressing heart condition. Mason was not the easiest person for anyone to deal with, including the wife he adored.

I was with Mason's wife when she died, and at that time Mason became an even angrier old man. Since he had no friends (for a good reason), we took him into our family. I made sure he had a place to go for holiday dinners and birthdays, and occasionally I would take him out to lunch. His behavior continued to be unpredictable.

About a year after his wife died, Mason became ill and had to be hospitalized. He told the medical staff he wanted to die and was angry about being forced to have transfusions. At that time, I no longer worked at the hospital. On the last day he was alive, which happened to be my birthday, I visited him.

Although Mason was still very angry, finally on that last day of his earthly life, he became open and receptive. He was willing to listen intently to the things of Spirit. As we spoke of love, he gained hope of reuniting soon with the love of his life. For the first time, possibly ever, I actually saw him release anger and move into peace. He died that night.

It takes some people longer than others to open and become receptive to spirit. Think of how Mason's life could have been different if he had focused on the love connection with his wife rather than hating the world.

Principle - Centered Living – Living a Balanced Life

If we ask questions about our life's purpose and moment by moment, move ahead expecting guidance, we will have it. Enduring a dark night requires faith, prayer and if possible, a spiritual director. Keep a journal and trust that all dark nights have a light at the end. Look for it.

One doesn't have to embrace the Christian religion or any other religion for that matter, to point to The Eight Beatitudes of Jesus found in the New Testament as a simple, elegant, all-inclusive doctrine that describes the principled or balanced life:

1. Blessed are the poor in spirit, for theirs is the kingdom of heaven.
2. Blessed are they who mourn, for they shall be comforted.
3. Blessed are the meek, for they shall inherit the earth.
4. Blessed are they who hunger and thirst for righteousness, for they shall be satisfied.
5. Blessed are the merciful, for they shall obtain mercy.
6. Blessed are the pure of heart, for they shall see God.
7. Blessed are the peacemakers, for they shall be called children of God.
8. Blessed are they who are persecuted for the sake of righteousness, for theirs is the kingdom of heaven.[74]

Many spiritual teachers and leaders of other faiths have also eloquently defined the principled life. In the workshops I've given, I've found that teaching a doctrine of living by principle is easier than teaching a doctrine of living by rules. **Simple principles, such as placing people before things and relationships before achievements, take all the complexity out of decision making.**

> *If it is true that Divine Energy runs through your being, do you see an aspect of self that is a co-creator with the Divine? That means you can trust that you are never alone. In addition, you expand God's creations through your creative expression.*

---

[74] *Gospel of St. Matthew 5:3-10*

In evaluating your life, you may find that in addition to meditation, making time for weekly reflection may be valuable. Take an hour or so for "re-creation," maybe on Sunday. You will discover that it is actually more refreshing to take an hour to balance and purposefully create, rather than engage in typical recreational activities.

You can use formerly "mindless" time driving, watering the lawn, doing dishes, etc., to quiet your body and listen to your deeper self.

Sacred Missions

Have you ever met anyone who seemed to know exactly what they wanted to do with their life from the time they were very small?

Some people may be dedicated to service, to make this world a better place. They may love people, animals, the earth or plant life. They could choose to become a social worker, psychologist, veterinarian, geologist, nurse, lawyer, agronomist, etc. Possibly they love archeology and enjoy digging for artifacts that tell the story of earlier civilizations. Maybe they love to teach; nothing gives them greater pleasure than inspiring young people or getting positive feedback from students.

Others may be equally dedicated to undoing all the good in the world by creating as much drama and chaos as possible. Somewhere inside of them is a ball of anger from which they hurl out massive amounts of hatred and resentment toward others and the world at large. Nothing is ever right; they are hostile, destructive and often deliberately cruel toward others.

How can we grow from our awareness of these people? Did they show up in our lives to teach us something we need to learn?

For a human who at one time was simply a heartbeat, what would it take to turn that tiny heartbeat into an individual who now harms others? From our perspective, obviously they are disconnected from their Divine selves and from a sacred mission that lurks in the shadows of their soul. Yet we are never in a position to judge others. Our perspective is too limited for us to claim to see the bigger picture behind other people's agendas and behaviors that appear subversive. If nothing else, observation of diverse behaviors can help us sort out what we don't want for ourselves.

Do not judge others, **period**. You have not walked in another's shoes. Ask yourself: what would it take for someone who was once just a perfect little being to act in the way they do today? What kind of programming was downloaded into that pure spirit?

I personally I had no idea exactly what I wanted to do as an adult when I was a young child. I did know I wanted to do something that would help people. I wanted a safe and secure future, meaning work, savings, and one day, children.

After my mother died, I ran across some interesting papers she had saved that shed some light on this idea of inner knowing at a young age. Among the papers were several stories written by my brothers and sister during the time they were in second or third grade. Apparently, since we all went to the same school, we had similar assignments.

My brother, who later was to become a pilot, wrote a detailed story about how he built a helicopter that was powered by his bicycle. He could lift himself up from the back yard and fly directly to school, saving three blocks of walking! A career as a pilot seemed like a logical path for him.

Another brother at a very young age wrote about wanting to be an adventurer, travel the world and have fun. He moved to Maui, became a ship's captain and now travels the world. He's surely had a lot of fun!

A third brother wrote a story about a singing train that carried children and toys to another town. The train broke down and everyone was sad. Who came to the rescue but an electrician?! You can guess what career this brother chose!

The fourth impressive sibling story is about my sister; she was the fourth of seven children. On the first day of first grade, she came home and announced to the family that she was going to be a doctor, and that is exactly what she became!

These stories were very revealing. In choosing their life work, none of the seven children in the family followed in the footsteps of their parents or relatives. In fact, I don't even remember any close family friends who were pilots, captains, electricians or doctors.

At a very young age, those children must have already had a deep awareness about their future life.

How about you? Was there a deeper part of you as a child that had knowingness about your future direction when you were a child?

The Role of Religion

What is the purpose of organized religion, i.e., religious bodies of worship that require a commitment to attend worship services, enroll in classes and participate in numerous weekly activities? Does this affiliation and active participation contribute to our soul growth? I believe it surely can, with certain caveats.

Religious organizations that condemn non-members or outsiders—those who are affiliated with other religious groups, agnostics, atheists, spiritualists, etc., provide a fertile breeding ground for bigotry, hatred and separation. They are an obstruction to the growth of any society. Religions that control their membership by preaching dogma that instills fear, threat, guilt and punishment, can be equally deleterious. Fear-based thinking embedded in the subconscious is one of the biggest obstructions to soul growth.

Likewise, any teaching that fosters codependency, i.e., obeisance and allegiance to other individuals, limits the mind from receiving guidance from one's Divine source.

All of the great teachers of history speak of peace, love, and service. In his book, *My Spiritual Autobiography,* the Dalai Lama writes: "I am convinced it is always more satisfying to preserve the religion of one's ancestors. They all recommend an inner transformation of our stream of consciousness, which will make us more devout people."[75]

If we are affiliated with a certain religion, in addition to remaining committed to our beliefs, it is important to keep an open mind toward beliefs that differ from ours. Pope Francis has been a wonderful ambassador to the world, speaking of love, compassion, tolerance, service, mercy and the uniting of people of all religions coming together for the common good.

A recent study in the *American Journal of Epidemiology* by the London School of Economics and Erasmus University Medical Center in the Netherlands found that the secret to happiness lies in sustained participation in religion. I would question whether it is religious participation or a focus on spirituality, service and community that is the reason for an individual's sustained happiness. What do you think?

Questions & Reflections

1. Do you attend church services for fear of condemnation or punishment if you don't attend?

2. Do you think you are acquiring special rewards ("brownie points") for attendance?

---

[75] http://www.amazon.com/Spiritual-Autobiography-Dalai-Lama-XIV/dp/1846042410, Dalai Lama. *My Spiritual Autobiography*, Rider, 2010, p.81.

3. If you go to church, do the services fill your heart? Are they uplifting and spiritually nourishing?

4. Through your religion, are there opportunities for you to serve others?

5. Do the rules of your religion keep you from doing what is most important at that time? Do regulations come before people, meetings, and service? (If so, whom are you serving?)

I believe that sustained happiness is created by a focus on our eternal nature. If religion supports this focus for you, then by all means, participate as a member of your chosen faith.

God Loves Us

I propose that God, Source, or The Divine, loves us and moves through our hearts. We are wise to look into our own hearts and see ourselves as an extension of that Divine love. The recognition that the Creative Source or God energy flows through us, helps us recognize our co-creative role in our life. Our soul-centered lives will then draw us forward on our soul path.

Life Is an Adventure - From the Teaching of Helen

If my mother, a beautiful, fun-loving, brilliant Irish woman, had a motto in her life, it was "Life is an adventure."

My mother found joy and excitement in even the simplest of things. She created adventure in her everyday life and enjoyed vicariously the adventures of her seven children. She even found a way to make the catastrophes sound delightful. If she had a central theme that brought her joy, it was playful creativity. To Helen, every part of life was an adventure.

More Questions & Reflections

1. If you had a spiritual theme to your life, what would it be?

2. Can you cite "before" and "after" examples in your life that have changed radically since your childhood days? How did these changes influence your life?

3. How is your lifestyle different from your parents' lifestyle? What factors do you think caused these changes to occur? Do you feel these changes are improvements or would you prefer to return to your parents' lifestyle if that were possible? Explain.

4. Is your work "just what you do" or does it have special significance for you? Explain.

5. If you chose to embark on a lifestyle that is different from your traditional group of people (family, religious group, etc.), did you encounter opposition? If so, how did you meet this challenge? Explain.

Exercise

Take several minutes for this exercise. Read it through first and then move through it slowly.

Begin your focus with slow, deliberate, easy breathing. Take several minutes to draw your awareness inward, and release from your body all distracting thoughts. Release your given roles, jobs, and expectations of others. Become totally present with yourself in this moment only.

Imagine your heart ever so gently desiring to lead you forward. You may feel as if your chest is opening to all possibility that is before you. As you imagine actually opening your chest, feel your heart gently leading the way, making your path easy. You may feel as if you are moving toward a beautiful light on the horizon. Because the deepest part of your heart already has all the information and answers you need, you can trust it and allow your soul self to emerge toward the light in front of you. You may feel as if you are being drawn toward that light.

Draw your attention inward, into your body and notice that your heart has its own light source. You may feel your heart light glowing and brightly illuminating the path before you. The immense, grand light on the horizon is like the sun, drawing you forward. Your own heart light is brightened by continually moving closer to that light source.

As you connect with Source energy, you feel drawn along your path almost automatically, as if there is a magnetic attraction. Life may start to feel like you are flowing with ease, as if smoothly floating down a river.

Remember this feeling as you are making life decisions. Note how your heart feels about your decisions. If it feels calm, peaceful, even effortless, you have initiated your co-creative flow and can move forward with ease.

Whenever you run ideas through your mind and anxiety and angst arise, try adjusting your course.

10

Why Do We Choose the Hard Way?

We can do this the easy way or the hard way. Why would we choose the hard way?

When we have a choice between trying something new and different or doing what is familiar—even if we already know the outcome will cause discomfort or pain, even if we have been told the new way will be (easier, faster, more successful, lead to greater happiness, etc.)—generally we will choose whatever is most familiar.

Fear of risk, of uncertain outcomes or just having to adapt to something new, is part of our earlier subconscious programming. The powerful scripts, such as: "We've always done it this way," or: "This is the only way it can be done," make our choices for us unless we consciously decide to release them.

As soon as you recognize that you are willing to open your mind and heart and look deeper into yourself, to begin a path of never-ending discovery, you will have taken the first step on the path toward personal growth and empowerment.

In your mind's eye, *actually seeing yourself on a path* can assist you in making decisions in your day to day life.

Some of us may seem to have easier paths than others. After we discover more about the challenges of other people, however, often we realize this was merely our limited perception at the time. **In reality, we can never judge another's path.** What may be a challenge for others may not be a challenge for us.

Conscious awareness that you are on an *eternal soul path* can help you transform challenges into growth.

If you were walking down a path in a lovely garden and found a big rock blocking your way, what would you do? Maybe you would turn around, quit and go back to where you were before. It might look like too much hard work to remove that rock. And then, even if you did remove it, do you really know what's on the other side of that rock? Of course not. The rock is obstructing your view.

Others with determination and curiosity who realize the importance of forward progress might walk around the rock or crawl over it. An engineer might build a skywalk over the rock, ensuring that others on the same path would have an easier journey.

We tend to gravitate and make choices toward what is familiar. It may benefit us to move from our comfort zone, from what is most familiar, to the edge of discomfort to learn a better way. The new way then becomes familiar and comfortable for us. **Learning leads to growth, additional knowledge and experience.**

The Hidden Secret behind the Sweets

"Sally" came into my office complaining of a sweet addiction. She was about 100 pounds overweight and she believed her addiction to chocolate-covered macadamia nuts was a simple sugar addiction. It caused her to hold onto 100 pounds or more through adolescence and well into adulthood; I suspected it was even more than 100 pounds.

In recalling her family history, Sally remembered almost nothing of her childhood. She did remember jumping for joy at age 12, when her father, a truck driver, killed himself. She wasn't sure why her father's suicide made her rejoice. She had very little recollection of anything else from her early years.

As we peeled through layers in her psyche, Sally was shocked to have memories of sexual abuse going back to very young years. After our sessions in a full waking state, she fought those bits of truth, thinking she was making it up or that it was impossible "for anyone to do that" to a child.

In one session, she recalled making a firm decision as a child: "When I become big, no one is going to hurt me." True to her plan, when she grew up she got big. Her body translated "big" as in "safe," to mean "physically big." No amount of dieting could override her subconscious need to be big in order to be safe.

In our healing work, Sally spontaneously regressed to pre-birth. As we cleared her birth issues we reframed her birth as a healing experience.

Sally's healing was amazing. Eventually she trained herself to work with children who had been abused. She changed her name and began a new life.

Clearly what seemed like a case of "I want to lose weight" was much deeper than that. Sally's courage to learn a new way brought deep healing and growth.

Fortunately, most of us will never deal with situations as extreme as Sally's. Nevertheless, allowing our seemingly physical or emotional imbalance to act as a red flag in getting our attention can serve to direct us on our authentic soul path.

The Smoking Cessation Therapist Who Couldn't Stop Smoking

"Gary," a 42-year-old male, came into my office to stop smoking. He was very health conscious and annoyed at his inability, up until this time, to stop smoking.

Gary had actually been trained as a hypnotherapist. He had tried many times to quit smoking himself but was unsuccessful. Being a smoker was embarrassing to him because he helped other people quit smoking!

Often, smoking cessation is a simple one-session 100% fix. If someone is still smoking or even thinking about it after one session, I know there's another subconscious obstruction that needs attention.

During the second session, we found a block which we removed. Gary was down to a couple of cigarettes a day but that wasn't good enough. Our goal was to be 100% smoke-free. In the third session, we dropped deep into the hypnotic state and regressed to the time Gary reached for the first cigarette. We then backtracked in time, one simple frame at a time, to see what led to that habit.

In the regression, Gary was shocked and confused to find himself on a golf course with his former wife. They had been divorced for a decade. As he relived the scene on the golf course, he remembered his former wife telling him she wanted a divorce. The announcement was unexpected. Gary was shattered and said out loud to her, "If you leave me I'm going to die."

Immediately he walked off the golf course and started smoking. He had no idea that he was following through on the subconscious message he gave himself, i.e., to die, one cigarette at a time. Since the shocking news of divorce put him in a state of overload, his subconscious mind was wide open and he took his own suggestion.

We reframed that message and Gary walked away from smoking cigarettes effortlessly, forever.

This is a very interesting case in that Gary gave himself the suggestion that he wanted to die. At the moment he said it, he felt like that. He acted on his momentary desire by engaging in a habit that could kill him one cigarette at a time. Now it was many years later. He had moved on in his relationships and he no longer wanted to die, but he was still acting on a false message.

Getting rid of cigarettes for Gary was the beginning of reclaiming his life.

I Do Not Want to Be Here

Dr. Z came to my office because he said he wanted to start exercising, but it didn't take long for me to realize that what he really wanted to work on were soul issues.

He was stressed, frustrated, and basically depressed. Somewhere in his memory, he recalled not wanting to be born into this life. He said he felt like he was pushed here, kicking and screaming. He had even used drugs in his earlier years to relieve his pain.

Dr. Z was a well-educated man who made a point of staying current on global issues, particularly political, financial and ethical ones. He had a lovely wife and several daughters whom he thought should make him happy. He also had a new job with a great medical group but none of this seemed to give him the happiness he truly desired.

The work began with the establishment of an exercise routine, which is the issue that had initially brought Dr. Z into my office. As he kept to his program, he started looking and feeling better. Eventually, this first step led to Dr. Z's transformational journey. It was a beautiful transition to watch. Together we reframed his dismal outlook of the healthcare system and he began to work from his heart as a compassionate healer. He also learned to find more joy in life and with his family.

As Dr. Z started to approach his life from a soul-centered perspective, everything shifted. His reconnection to soul changed his outlook on life, family and his profession.

Questions & Reflections

1. Do you feel stuck in a job or situation?

2. What would you like to be doing instead?

3. Name the obstacles that you believe are holding you back.

4. Do you have a plan for removing those obstacles? Explain.

Exercise

What thoughts about your future fill you with passion and excitement? Can you ask for spiritual assistance to help you create situations that better serve you and bring joy? Of course you can. Here's how to do it:

1. Recognize that something is not working for you.

2. Look into your heart and see how it feels regarding the person, place or situation.

3. Begin to imagine a variety of scenarios as if they already exist. Note how your heart responds to them.

4. Ask for spiritual assistance.

5. Keep your mind open and clear. Look for synchronicities and meaningful symbols to give you direction.

Often information is delivered to us in very subtle ways. Pay attention. Your future is created by your thoughts.

Visualization

If you feel stuck somewhere in your life and you want to release the block that is holding you back, you can explore the possibility of using visualization. Remember, your mind cannot tell the difference between fantasy and reality. **As you create a picture in your mind of what you currently feel like, you can change the picture you are imagining and change your feelings.** If you feel stuck as if in mud or down in a hole, you can imagine pulling one foot at a time out of the thick mud, or you can visualize a golden rope to lift you up and pull you out of the hole. Your mind will create the perfect symbol for you.

By now you are practiced in using a breathing technique for quieting your mind and body. After several minutes of quiet focused breathing, ask yourself if there is any area of your life where you feel stuck. It could be holding you back from advancing in your job or from trying something new. It could even be causing you to hold onto anger that interferes with your peace.

How does the stuck feeling express itself in your body? Does your stomach feel nervous, do you feel heaviness in your chest, are you afraid to speak your truth, or do you just not feel the joy you desire?

This feeling and its related issue may be holding you back from your soul's evolution. You can shift the feeling by finding a different image that allows you to move past your block.

Imagine putting a picture in your mind that is a visual for the feeling of being stuck. You might imagine that someone has placed a hand on your chest to literally hold you back. You could see a big rock on your path, or maybe you're stuck in a hole and can't get out. You could even imagine your feet being stuck in quicksand or your throat tightly closed up.

Choose a picture that best represents your feeling of being stuck. Now imagine this picture as your reality. What could you do to change the picture? Could you take a step backward and walk around the person whose hand is on your chest, holding you back? Could you walk around the rock or build a bridge over it?

Could you look to the top of the hole you are in, see a light and find a hand reaching down to help you? Could you pull one foot out of the quicksand, and then the other? Could you open your throat the way an aperture of a camera opens, to allow more light to enter so that you could express yourself clearly with words?

Ask your wise subconscious to show you the ways to literally remove the obstacle, step by step. Then open your mind and eyes to the synchronicities and opportunities that will come your way from the desire you planted in your subconscious mind.

11

Challenges Become Gifts

When we are in fear we are focusing on what we don't want.

Am I Still Connected?

We all have "Dark Nights of the Soul," those times when we feel completely defeated and hopeless, wondering if we really are connected to Source or God. Why have we been abandoned?

Starlight

Earlier I mentioned one of my own "dark nights of the soul" that occurred during the years when I was raising six children as a single parent. On another occasion, once again I was feeling confused and completely overwhelmed. Added to my responsibility for six children was the housing of a neighbor's child whose father traveled a great deal. Four of the children were healthy robust 13-year-olds. I felt outnumbered by 7 to 1.

One night I went out for a walk after the children were finally in bed. I had been meditating and praying for support but felt so very alone. In walking up a long hill back to my home, I asked God, my Source, for strength: "If you know I am here, I need a sign," I said out loud. I was so over my head in responsibility and I felt my faith was being severely tested.

As I looked up at the night sky I saw a brief flash of a shooting star. I live close to the ocean and on the edge of town, where fog and too much ambient light usually make it difficult to see any stars, much less a shooting star.

I questioned what I had seen. Maybe it was just a moth that flew into the path of a street light. While I was contemplating this, I saw another shooting star. I was still slightly skeptical, especially since the second sighting happened so quickly after the first.

And then, what did I see? A third shooting star! This one was larger and visible for several seconds. Now I had no doubt about what I was seeing.

Although I was still feeling alone, I was overcome with awe, knowing that God, my Source and support, was aware of me and giving me the sign that I had asked for. While I was offering profuse thanks, actually feeling a little bit embarrassed about asking for a sign, a fourth star shot across the sky overhead.

As I entered the house that night I was still not sure how I was going to manage the activities of the next day, but I did know with absolute certainty that I wasn't doing it alone.

Blocks

At times, it is the very simple human things that block us from full awareness. Other times, the issues are larger and more complex. Typical blocks or obstructions are addictions, anger, fear, doubt and learned patterns that are not in alignment with our true selves. Any blockage of thought energy will challenge forward movement on our soul path.

If you were driving down a country road and a big brown cow was blocking the lane ahead of you, what would you do? Some might go around it, others might choose to honk and make the cow move, or even get out of the car and steer the cow back to a safe area behind the fence. Any of those actions would allow us to go forward on our way. Some of those actions you chose could not only clear your way and protect the cow, but also improve the path for those who follow behind you.

If all of our blocks were as simple as a cow or chicken blocking the road, life would be easy. But they are not!

Challenges Help Us See Things Differently

When working with cancer patients over time, it was common to hear many of them say that the cancer was the best thing to have happened to them. It changed their lives: their outlook, relationships and priorities. Most people would agree that cancer is a challenge no one wants to deal with.

A devastating illness or loss can help us make an internal shift in perspective. Nothing is ever the same once we realize our time is measured. The illness or catastrophe can allow us to look deeper within ourselves, to consider what really matters in the "bigger picture" of life. We can learn to let go of what is insignificant and embrace those qualities that are important. Illness or loss can also allow us to gain compassion for others. When we have personally experienced a major challenge, we know what it feels like and can more fully empathize with other people's pain and grief.

The more obstacles we overcome, the more experienced we become in navigating through the next obstacle, and the next one after that. Building on our successes, we learn to be very creative in moving energy through a wide variety of challenges. The practice of finding "soulutions" for our challenges becomes a major contributor to our soul growth.

Viewed symbolically, our life crises may tell us we need to break free of beliefs that no longer serve our personal development.

Strong as an Ox

"Katie" came into my office about 25 years ago seeking soul direction. She had just been diagnosed with severe osteoporosis at the relatively young age of 39. With young children, Katie didn't know what the future of her health would allow her to do. Her doctors didn't give her much hope for recovery and most of her medical questions remained unanswered.

Katie wanted to re-examine her lifestyle, asking, "If I can only be physically active for a short while in the future, what do I want to do differently?" In addition, she wanted to know why her body developed that disease and what she needed to do to heal.

The key to Katie's healing began with identifying the emotional part of the disease. What message was her body trying to tell her?

Using a regression technique, we went back to the time when Katie's bones were healthy. She saw where she was in her life and marveled at her strength and energy. We then progressed forward in time to the moment when her bones began to give up calcium and lose strength.

Katie recognized that throughout the years of her marriage she had given up her power and strength. She avoided standing up for herself. In her mind she saw a picture of a tall Weeping Willow tree whose branches were moving in a gentle breeze. She immediately knew what that meant. She had become like that Weeping Willow tree, bending and swaying to avoid contention. She recognized that her body took that emotional pattern literally and transferred it directly to the very bones that helped her stand upright.

This recognition was enormous for her!

I asked Katie if this was a pattern she wanted to continue. She said, "Absolutely not!"

I then asked her what she would like her bones to look like. In her mind, she saw a big, strong, solid white hip bone of an ox sitting on a mountainside. The hip bone was still strong, long after the ox had departed. "That is how strong I want my bones to be," she said.

"What do you need to do so your bones can reclaim their strength?" I asked her.

"I need to speak my truth, nourish my bones and reclaim my power," said Katie. "I'm going to do something I have always wanted to do—I'm going to teach my children to ski." And she did! Each year her bones progressively gained strength and twenty years later she informed me that she had her first normal bone density scan!

Perspective – It's All in How We See It

We see things not as they are but as we are. The world mirrors back to us the basic truth about ourselves.

Perspective in all things big and small can keep us focused on what is most important. Recognition that some things we stress about today won't even matter next week can help us put them in proper perspective.

I like to use the term "inside eyes" when referring to a part of our vision that can look inside ourselves and read our feelings. Inside eyes can also be used to look at other individuals and see what is beneath the surface of their actions. For example, what is the reason for a person's unusual or aggressive behavior? When we view it from an inner perspective, we may gain important insights that will allow us to respond to that individual's behavior with compassion and understanding rather than anger or resentment.

We can use our inside eyes to look within ourselves and gain insight just like Katie did, to strengthen her bones.

Express Yourself

Health is not just a matter of thinking happy thoughts. Sometimes expressing and releasing a burden that has been suppressed for many years may jumpstart the immune system.

Dr. David Spiegel, Co-chair of the Department of Psychiatry and Behavioral Sciences of the Stanford University Medical Center, successfully demonstrated that the ability to express emotions such as anger and grief can improve survival rates in cancer patients. "It's healthy to feel anger or sadness," says Spiegel. "Women who check out by trying to constantly control their emotions are under more stress."[76]

Since emotional expression is tied to a specific flow of peptides in the body, chronic suppression of emotion results in massive disturbance to the psychosomatic system. Freud describes depression as anger directed toward oneself. Humans share the molecules of emotion with the most modest of creatures, even a one-celled being; thus, there is unity in all of life. Dr. Candace Pert claims that all emotions are healthy because emotions are what unite the mind and the body.[77] I tell my clients that the only unhealthy emotions are the ones that we store inside rather than allowing ourselves to express them.

Addictions and Emotional Pain Can Block Our Growth

Caring for both our physical and emotional bodies will automatically affect the health and well-being of our soul body. **Mental and emotional clarity as well as keen sensitivity, are basic to our soul growth.** Addictive behaviors and other types of emotional numbing can lead to chronic obstruction. We find ourselves investing large amounts of energy to medicate the craving and withdrawal symptoms common to any addiction.

---

[76] http://www.shlnews.org/?p=55

[77] Op. Cit., Pert, Candace. *Molecules of Emotion: The Science Behind Mind-Body Medicine*, p. 192.

Denial or the suppression of anger, resentment, fear, jealousy and other forms of emotional pain is another energy thief.

It may be idealistic to think we can release an addiction or solve a deep-seated issue of grief by ourselves. After acknowledging we have a problem that we sincerely wish to resolve, it is vitally important to seek help as soon as possible. Sometimes these behaviors may continue to hold us hostage without our even being aware of them until serious damage has already led to illness, loss of valued relationships, a job or career opportunity, etc.

The "upside" is that we can always turn around a challenging situation by removing the obstruction and once more creating free energy flow for a pure connection to the soul. Trusting the soul and allowing it to be in charge of our decisions is an important step in the process of releasing blocks.

What Keeps Us Stuck in Our Comfort Zone?

Habits or behaviors and actions, belief systems, stories that prescribe the best way to live, and expectations, are examples of programs that keep us stuck in our comfort zone. These subconscious stories or energy patterns continue to deliver messages to the conscious mind whenever the right words, sounds or symbols trigger or deliver the "software" code for that program. We are conditioned to act according to our stored programs.

Even before birth, we begin to store programs in our subconscious mind-body through our mother's chemistry. Originally, the primary reason for this programming was to keep us safe. It was developed for survival purposes, as a fight or flight mechanism. During our life we can be so distracted by the stories we tell and the roles we play, we may never learn who we really are.

Group programming or "the group mind" was one of the earliest forms of survival. It is still alive and active today. Tribes, cults, religions and all other groups that bind people to a code of beliefs and rituals or practices, are examples of group programming.

Carl Jung said group mind is the lowest form of consciousness because individuals involved in a negative (destructive, ill-meaning) group action rarely accept responsibility for their personal action.[78]

It is a big step forward for us to become aware that we create our reality and then proactively accept responsibility for our behaviors and actions. Transformation occurs when we are willing to risk change and move out of our comfort zone. Sometimes that willingness is founded in desperation, fear, anger, grief, and a host of other stressful emotions. Whatever the motivation, once we determine to change our current situation and "make the shift"—once we open up to our Higher or Inner Self through a heart and soul connection—we start to experience positive, life-affirming results.

Judgment Places Obstacles in Our Path

What is it like to be me when I am not judging myself?

What would your life be like if you judged yourself less? What would it feel like to experience interaction with others from a place of non-judgment?

How much kinder would your heart feel if you viewed the behavior and actions of yourself and others through love and compassion rather than judgment?

It is not uncommon for me to see clients who are filled with self-criticism. I remind them they are doing the best they can, based on their subconscious beliefs. The world can seem even tougher when we are criticizing ourselves rather than looking compassionately within, to the loving heart.

---

[78] http://www.solhaam.org/articles/mind1.html

Judging others may lead to comparing them to ourselves. Since our view of others is so limited, this is unrealistic and performs a disservice to ourselves and to them.

Overcoming Fear of Expanding Consciousness

Fear can be one of the biggest obstacles to soul growth. It can also prompt us to make poor decisions. In a state of fear, our consciousness works against us. *It is actually focused on the opposite of what we want.* Focus on fear has devastating repercussions, both emotionally and physiologically.

Fear acts as an invisible barrier to creating what we want. It also shuts down our intuition. We remember failure from the past programs running in our subconscious. Any similar situation or experience triggers fear from the past. We fear moving forward and away from old memories that we are programmed to believe will protect us—yet do we really know this? Of course not! Blocked from our intuition, we have no way of confirming the validity of our old programs.

Without our PGS (Personal Guidance System) giving us instructions for staying on our soul path, we can easily lose our way through the labyrinth of "what if's" and worst case scenarios that crowd into our consciousness. The whole world looks gloomy and gray when we lose our direction and connection to soulutions.

Most of the worst decisions I've personally ever made were based on fear. Take a moment and think of the decisions you made when you were fearful. How did they work out for you?

The reality is that most of the stuff we worry about is temporary or projected into a future that hasn't happened yet. Often what we fear will *never* happen. Even if the worst-case scenario we imagined played out, emotionally we would have put ourselves through the ringer not once, but twice!!

Ask yourself about the situation: Will this matter in 10 years? Will this matter in five years? Will this matter in one year? Will it matter tomorrow?

It's amazing how quickly we can put things into perspective.

The good part about any type of blockage is that when we place it in perspective, it can bring us to the awareness that we have the ability to shift our focus to what we *really* want.

Ask yourself what you really want and then *feel it* as if you already have it.

What do you want that will matter in ten years—or even a hundred years? What do you want in terms of expanded consciousness that will matter to you eternally?

I made a decision early in life that when I had an adversity, in addition to moving through it for myself, I wanted to make the road easier for others who had similar challenges. Maybe that decision came from being the oldest child of seven and being told that I must be a good example for my siblings.

No one formally taught me this principle of making the path smoother for others; it just felt like the right thing to do. Little did I know that this firm decision would open doors for me that would greatly expand my soul purpose.

I Want to Help Others

In 1983 after giving birth to quadruplets, I lost one of my babies eight days later. Ryan had actually been born healthy. The doctors thought he needed a minor procedure common to premature babies. When the procedure was done, some blood vessels were accidentally cut and he died three days later. It was a devastating loss.

I still had three surviving babies and a two-year-old to care for as well as the responsibility of moving my family to a new location 1000 miles away from family support.

I was well versed in grief counseling from my work as a nurse. Yet in the 24-hour daily demands of caring for three preemies, it was a challenge to find time to grieve. I didn't want anyone else to ever have to go through what I was experiencing. It didn't follow a typical grief reaction because I was dealing simultaneously with joy for my surviving children. Also, I was exhausted! I was hearing comments from others, such as: "Three out of four isn't bad"; or, "What would you have done with another child anyway?" Those are NOT comforting things to say to a grieving mother! I recognized that others simply didn't know how to react and they were trying to be comforting.

I used the long hours of the night while caring for babies to process my emotions. I discovered this was a unique form of grief. I had never seen any material or data about managing grief under these circumstances.

A multi-fetal situation produces unique aspects of grief that are profoundly different from other situations involving loss. I quickly learned that I would always have reminders of my lost child and lost unit when people began to call my three surviving quadruplets "triplets." I realized among other things that I was so busy and tired, the natural tendency would be to just move ahead without fully grieving. I knew this was unhealthy.

During that time, I used the dark hours of the night to meditate and gain understanding. Two years later I began a research project in which I interviewed over 100 women who had lost one or more of a multiple birth unit. I started a support group called "Tender Hearts" and wrote a white paper to educate parents and medical professionals in how to manage this situation of simultaneous rejoicing and grieving. I found that often parents were so tired and overwhelmed, they ignored the loss. Medical professionals had no idea how to manage these situations.

Shortly after I wrote the paper, I was surprised to be contacted by a physician from Chicago who was the head of International Society of Twin and Multiple Birth Studies for the United States. He asked me to be the keynote speaker in Amsterdam that next year for an international conference of physicians and researchers engaged in the study of multiple births.

The entire experience was humbling. At that time, I had only an RN and a Bachelor's degree and the group of individuals to whom I would be presenting had multiple M.D. and Ph.D. degrees. An extremely kind and gracious group, they welcomed the information I presented.

What started out as a simple desire to understand my unique situation and help others encountering similar circumstances ended up being a paper that was published internationally in a scientific journal with far reaching effects! The achievement was so much greater than I could have imagined. I got the information out to the world to those health professionals who could spread it further.

All of us have times, places and situations that are unique to us. Each of us can make a difference to someone, somewhere, with our individual experience and perspective. Sometimes the little things are the big things. We may never even know what we may do or say that might lead to a change or growth for someone else.

We Must Be Willing to Risk Discomfort

If we want to effect change in our lives, we must be willing to stretch beyond our comfort zone and take risks. **Feel the uncertainty about making any change—and do it anyway!**

Too much discomfort leads to panic, which blocks our ability to move forward. We don't want this to happen. What we do want is just enough discomfort to allow us to challenge old patterns for the purpose of replacing them with ones that better serve us.

To create a new comfort zone, move out of your current zone by easing yourself forward to the edge of discomfort until it feels comfortable. Too big a push forward can trigger fear which may block you. There's no need to rush the process. This is not a race. Moving to the edge of discomfort is more like steadily accelerating forward and paying increased attention. **We want to live on the edge of discomfort and not in the midst of it so we can keep growing.**

Make the Shift!

If you will think of moving away from old outdated patterns of the past toward patterns that work better for you as *a shift in consciousness vs. a change in direction,* it will feel more effortless.

Shifting our thoughts and actions just a few degrees at a time will allow us to arrive at a very different destination. You could compare this to setting course for navigating a ship on the ocean. If you started on the east or west coast of the U.S. and adjusted your course just a few degrees, your destination might be a different country or even a different continent.

What new ideas can you create to bring ease into your journey as you progress forward?

The 'Not Good Enoughs'

My most recent book, *An Everyday Guide to Joy & Abundance: A New Approach to Living with Ease,* is a collaborative project with Beth Wonson, who is an executive coach and business consultant.

When Beth came to me for help, at once I realized she was a very bright, gifted woman who had become disconnected from her authentic self. All the clutter in her life was holding her back from moving forward on her soul journey.

In the book, I describe the tools I used as a therapist to help Beth understand that much of the pain she had created and participated in throughout her life was a result of the loss of her deepest desires. Her subconscious programming was everyone else's but hers. She knew exactly what to do every day from the moment she woke up until it was time to go to bed. Her life was all mapped out for her, or so she thought, until suddenly everything started to fall apart.

Together we asked the big questions: What did she really desire? Who was she? Why was she here? Did she have a larger purpose beyond her 9 to 5 routine, beyond a highly stressful lifestyle that didn't seem to be going anywhere except around the clock to the next day and the next?

We set to work and started to clear out the blocks or "rocks," as Beth calls them, which were obstructing her soul's authentic journey. As soon she began to release the programs that she had never authentically resonated with—all the "should's" and "must do's" that had become deeply embedded in her subconscious— Beth started to create her own colorful palette of exciting soulution-based options. The remarkable transformation that occurred really did deserve a book.

Many people are highly self-motivated like Beth. All they need are encouragement and a few tools to initiate the process of reconnecting with the essence of who they really are.

Joy and Peace Come from a Much Deeper Place than Pleasure

Joy and Peace, emotions of the heart and soul, can be experienced even if the body is ravaged by disease or the mind is consumed by grief. All of our daily struggles—finances, weather, car problems, traffic, annoyances, difficult people—can be used as opportunities to stretch our soul. In this light, every moment of life becomes even more precious.

Release Your Baggage

What is baggage? One of the advantages of the fairly recent rules when traveling by air is that we carefully pack our bags with only what we need. We now have weight and bag limits with our travel. Before those weight limit days, you could pack your suitcase literally full of rocks and usually no questions would be asked.

Genny's Rocks

My dear cousin Genny had come out to California for a visit to help me with my daughter's wedding. After the wedding we spent several days leisurely walking on the various beaches in the area. Genny loved to collect beach rocks. There are no beaches back in Minnesota where she lives. We took an empty bag with us each day and she would find unique and beautiful rocks not only for herself but for her two daughters, eight sisters and sisters-in-laws as well. By the end of the week she had quite a collection. We packed them all in a medium-size bag to ship back to the Midwest.

When it was time to fly home, I took her to our small local airport and walked with her to the check-in counter. As the man at the counter reached over to grab her bag, assuming it was a normal weight, he was shocked. He could scarcely lift the bag. He asked her in an annoyed tone, "What do you have in there, lady—ROCKS?"

We both burst out laughing. The man did NOT think the heavy bag was funny! Genny responded politely, "Yes—rocks!" I'm not sure whether or not the man behind the counter believed her. She did get her precious rocks on the plane to take home and share with her family.

It was actually the very next week that TSA came out with a weight limit on bags. Jokingly we told Genny it was because of her bag of rocks!

What kind of symbolic rocks do we carry with us? Stories, baggage, old beliefs that hold us back from moving with ease in our life?

Carrying unnecessary baggage is weighty. Baggage or rocks for us may be just old stories, or beliefs that no longer serve us. We may have learned these ideas as children. Use your imagination to recognize and examine the thoughts, stories and ideas that no longer serve you. Then let go of excess baggage and travel through life lightly.

Remember, when your mind thinks of something, your body creates the chemistry that makes it feel like your reality. Since our mind cannot tell the difference between fantasy and reality in terms of how we physiologically react, our mind and thoughts can be a very powerful tool to assist us in removing the baggage.

As you spend time in quiet meditation requesting clarity, you might surprise yourself at the rocks or stories, the things you are holding onto, that you discover.

Regardless of the rocks or stories you find, the idea of holding onto something you no longer need can move you to action. As you release what you no longer need or want, you create new space to bring in what you *really do want*. This action can lead your emotional and soul body to a place of higher consciousness and freedom. You can then allow your deeper wiser soul-self to bring up a "soulution" and send it to the Consciousness Information Hub to help you get unstuck.

Here are the steps for achieving a soulution:
1. Stay calm.
2. Carefully examine the place where the block has occurred. Face the problem.
3. Consider your options for releasing the block. Use your imagination!
4. Set to work.

5. Be proactive and conscious of where you are in the process of removing the block.

Enjoy the feelings of lightness and freedom when the block is finally gone. You may also experience some unexpected surprises—good ones!

Questions & Reflections

1. Can you name some challenges you've overcome? How did they affect your life? What made you decide to take action and overcome them? Explain the actions you took. Were you satisfied with the outcome? Did the challenges help you see things differently? How? Explain.

2. Can you name current challenges you are working to overcome? Explain.

3. Are you currently working through issues that are keeping you from moving forward? Explain your plan of action.

4. Have you ever experienced discomfort when taking a risk? Explain. What was the outcome?

Visualization

We have mentioned before the power of the mind to change how we feel.

Begin focusing your mind and quieting your body by paying attention to each easy breath.

What do you feel like when you are being held back or blocked by an emotion? Identify in your body what it feels like. What does your heart feel like? What does your stomach feel like? What does your throat feel like (open or closed)? What do your feet feel like (stuck or free to run)? What does your head feel like?

What baggage are you holding onto that is causing you to feel this way?

Can you imagine standing at an airport baggage carousel and putting back bags and packages that you have collected that are not yours? Put back all of them, one at a time. Then the real owners can step up and claim them. You can reclaim what is truly yours that serves you. It may be one lightweight bag with wheels. In the bag you may have issues that belong to you that you can work on at a later time.

Another way to release old stories is to recognize that these stories do not belong to you. You can also place books of old stories on the airport carousel. Then you can create new and better stories for yourself that will allow you to live with ease in the moment.

Part III – Soulutions

Creating Your Heart Path to Wholeness

What is the purpose of our life except for soul growth?

12

Open Your Awareness

Becoming conscious is not easy. It requires
responsibility to act upon what we learn.

Our memories and attitudes are the rules by which we determine the quality of our life as well as the strength of our bonds with other people.

Consciousness gives us the ability to release old patterns or programs and embrace the new with an awareness that all things begin at the appropriate time. One aspect of becoming conscious is to **live in the present moment and appreciate each day.**

We can only open to awareness if we **commit to the conscious process of self-analysis and self-evaluation.** Evaluating our beliefs is a spiritual and biological necessity. Our physical bodies, minds, and spirits require new ideas to thrive.

Once we know the truth of something, it changes us. Becoming conscious, opening to our awareness, may mean changing the rules by which we live and the beliefs we maintain. The goal is to be able to fearlessly handle all changes in our mind-bodies while absorbing the message of truth contained in the change.

If we believe that our soul is eternal and connected to the Eternal Divine, all that we do in this life matters not only to us and others, but also to our planet, this earth plane, and beyond. Life becomes much more exciting than the next new house, vacation, job promotion, or retirement account.

Each of us as a unique soul has a particular spice to add to the soup of consciousness. Think of your authentic self as a special herb with its own savor. Every soul can make a difference!

How can we best develop and use our talents and gifts to make a difference first and foremost to ourselves and then to others, in order to contribute to our own soul's evolution?

How can we open our awareness and allow the Divine within and beyond to guide our path not just during our final days but throughout our life?

Awareness and acceptance of our authentic soul self comes first. How do we honor our authentic self? What does that mean?

We honor our authentic self when we do what's right for ourselves in spite of what others or society may expect of us. This may be a hard concept to grasp because many of us are taught that to put ourselves first is selfish.

In reality, when you make a decision based on what's right for you, at the soul level, it is the most *unselfish* thing you can do. The fact is, **when you do what is right for your soul, it is right for others. Doing what is right, always through the emotion of love,** is a guideline that can carry you through life and expand your capacity to love and serve.

Our Memory Bank

Is it possible that everything we need to know is already stored inside of us—already at our disposal, just for the asking? When we need information, is it possible that sometimes we can allow it to come to us?

The answer to both of these questions is yes. This is information we have never been formally taught. It is a deep knowing or part of our nature.

If this is so, how do we access this wisdom?

Learning how to expand our awareness is one of the most important components of Soul Science. Through this expanded awareness we are able to find soulutions to our challenges and move forward on our soul path with ease.

I have used many techniques to help others learn to use their own Personal Guidance System(PGS), as I like to call it. **Understanding how your mind works and the language it speaks is key to understanding and accessing everything you need.**

The first step toward achieving this understanding is to **learn how to quiet your mind and trust the subtle messages, thoughts, pictures, or promptings you are getting.** I can assure you, your deeper self is always trying to assist you. Sometimes the information is not logical, so we dismiss it.

As I was in the last stages of developing my previous book, *An Everyday Guide to Joy & Abundance: A New Approach to Living with Ease,* the front and back cover files arrived. I was delighted to see that they looked wonderful! The background color was the right shade of green, with very kind endorsements on the back cover, and my daughter's beautiful art was on the front cover. I showed the files to a couple of friends and emailed my editor, Carol: "PERFECT!" I was done!!!

The very next day as I began writing early in the morning, I heard a slight prompting, "Go back and look at the book covers again." It was a very, very small nudge. Logically I thought it was ridiculous, but I listened.

After six people had reviewed the covers, all of them had missed a glaring error: my name was misspelled on the front cover of the book! The 'r' was left out of Mary! How did we all miss that?! I was so grateful for the importance of listening to even very slight nudges.

A simple thought or feeling is one way our higher self communicates. There are other ways that we can access information on request.

Our deeper mind can often communicate through **symbols, metaphors, synchronicities, and numbers.** When you ask for information, pay attention. **Look for those indirect or subtle messages and answers.**

Heather, my daughter-in-law, was recently telling me about their search for a home. She said, "We love our location on the Central Coast (of California) but real estate prices are so much higher than average." She and my son had surveyed the surrounding areas and they found a town where they thought they could afford a home.

Heather's request to her soul was that if this was the right area, to let her know. If it was not the right area, she asked her soul to let her know this as well.

After numerous trips to this town, she saw signs that clearly it was not the right place for their family to buy a home. In broad daylight, they just happened upon two car break-ins and witnessed a drug deal, among other things. When my son saw a man riding a bike down the street wearing only a pink tutu and cowboy hat and waving something around—seemingly on drugs—they said, "NO, that's it! We do not want to have our children live in this neighborhood!"

The town where they were looking has many lovely areas and was known to be safe. Apparently for reasons unknown to them, the timing or location wasn't right. Heather got her answer, loud and clear because she asked and paid attention.

One year later, they found a great opportunity to live and work in a different area. They put an offer on a home and were ready to move in only six weeks after they first saw their new hometown.

How Can We Get More Direct Information?

When I work with clients, I help them access their subconscious mind in order to gain the information they're seeking.

If a picture is worth a thousand words, a metaphor is ten times more valuable. Our deeper subconscious mind communicates very effectively through the use of symbols.

As I have mentioned previously, often I use a metaphor of a lake or large body of water in order to talk with the subconscious mind. Everything above the water level represents conscious awareness. Below the water level and out of sight is an area that is loaded with every answer we need. These answers are delivered in terms of symbols or objects.

Often I will lead a client through a guided meditation into the state of hypnosis. Sometimes I have them imagine the water as a large, quiet lake. Other times, it is a deep well that runs to the center of the earth. In that state of active imagination, we bypass the logical mind and drop into the place of inner knowing. I will ask them to feel like they're reaching below the surface of the water with their hand.

I will direct a very specific question about the information we are requesting. I may ask for something to help with a decision that is being made or a next soul path step. I may even ask the mind to let us know if there is a block to healing. As the person is visualizing and feeling the temperature of the water on their hand, the perfect object always surfaces with ease. In my years of using this technique, rarely will anyone pull up the same object. If they do, the object may mean something entirely different.

Sometimes a direct answer is pulled up but more often, the object or symbol tells a story that contains the information requested. Other times that object or symbol needs to be questioned and examined carefully in order to discover the details.

Once an object is pulled up, I usually begin by asking if my client has ever seen this (the object or symbol) before—is it something old or new? If they say they have seen it before, I'll ask where and when. What would this object be used for? What would this object be trying to tell them about the answer they're seeking?

If the answer is still unclear, I deepen the trance, count to three and tell the person we are giving the object a voice. I then tell the client to ask the object why it appeared and what information it has for them. Immediately after the question is asked, I will say: "Quickly now, become the voice of the object and answer." We set up a two-way communication between the subconscious message and the conscious mind. This is a form of Empty Chair Dialogue in the Gestalt Therapy model.

It is fascinating to watch and participate in this. It is a very powerful way to communicate with our deeper self.

This type of communication can be taken even deeper, from a seemingly emotional issue to a deep place of soul knowing.

You may follow this line of questioning for yourself as you learn and practice dropping into a meditative state.

What Is Stopping Me?

"Jose," a middle-aged man came into my office with a strong sense that something was blocking him from success. He had been in the early technology boom and was very successful in many business ventures. Recently he made one bad business decision and since that time he had been stuck.

Copious research is available on the subject of entrepreneurial behaviors, specifically pertaining to entrepreneurs at the cutting-edge of new ideas and their ability to consistently spot opportunities.[79] I knew Jose fit into this category. His problem now was that he would make a decision, take action, second guess himself, and then pull out, recognizing later that he was correct in his first call. When he came to see me he was near panic, fearing bankruptcy.

I suggested to Jose that just making money might not be enough motivation for his soul. I asked, "What is the deeper reason you want success, more than just money"?

As Jose thought about that, he realized that what he truly wanted was to reclaim his intuitive self. He wanted his confidence back.

"That is a good enough reason," I said. "Let's go to work."

As I directed Jose to reach into his subconscious mind to find what was blocking his confidence, you can imagine our surprise when he pulled up from the depths of his subconscious... his mother!!! She was yelling something at him.

I asked him to listen to the words she was saying. He recalled that as a young man, his mother was very upset when he didn't go to college. She told him if he didn't go to college, he would never be successful; he would be broke for the rest of his life.

---

[79] "The Concept of 'Opportunity' in Entrepreneurship Research: Past Accomplishments and Future Challenges," by Jeremy C. Short, Texas Tech University; David J. Ketchen, Jr., Auburn University; Christopher L. Shook, R. Duane Ireland, Texas A&M University, http://jom.sagepub.com/content/36/1/40.abstract; "Opportunity Recognition by Successful Entrepreneurs - A Pilot Study," by Gerald E. Hills, University of Illinois at Chicago, https://fusionmx.babson.edu/entrep/fer/papers95/hills.htm

In spite of this, Jose was very successful until that first failure—which brought back his mother's message.

He addressed his mother by putting her image in a chair in front of him. Jose told her how he felt about what she said. He also took his power back by telling her of his regular business successes. Jose regained his confidence, which had been shattered by one bad decision that had brought back an old false message.

Expanded Perspective & Soul Guidance

After leading a client deep into a hypnotic state beginning with guided meditation into a beautiful garden, instead of using water as a metaphor for the subconscious mind, sometimes I move them, step by step, up the front of a magnificent pyramid-shaped mountain. We ascend all the way to the top to find a large shaded entrance to what looks like a cave. At the front of the cave is a heavy wooden door without a handle. As they approach, the door slowly opens.

This is a beautiful guided meditation. With practice, you can do it yourself. **The key is to focus, be clear about your intention and be willing to accept the information you receive.**

After walking into the dimly lit cave, the door closes and I give the client a few minutes to let their eyes adjust to the light. The cave has many nooks and crannies and travels deep into the mountain.

I then tell them a messenger or message will appear to give them some important information they need at this time. Over the years, my clients have produced a variety of messengers, from actual angels to Star Wars characters. The message may be a butterfly, or it is not uncommon for someone to be handed a note or envelope. Most of the time they completely understand the message.

When someone receives a note in a language they cannot read, I give the post-hypnotic suggestion that in their dreams that night, the rest of the message will clearly unfold.

After the message is received I remind the client to thank the messenger. They usually receive a nod or verbal, "You're welcome."

We then go back to the large wooden door at the entrance of the cave. As it opens to the outside, the sunlight is so brilliant the client may need a few moments to let their eyes re-adjust. They look out from the mountaintop and with expanded vision, everything appears brighter. Their perspective is enhanced and they hold the message they received deep in their heart. As they follow through and act on that message, they reconnect with their authentic soul path.

Questions & Reflections

1. What does it mean for you to "live in the present"? Give some examples in your life.

2. If you are now consciously living in the present, how has this changed your life? Give examples.

3. Do you have any skills that come naturally to you, i.e., as if you have a natural talent or knowledge of them? Explain.

4. Have you created a lifestyle that is centered around these natural skills or abilities? Explain.

5. Have you developed your own method for accessing your soul and receiving messages? Explain.

The key to expanding your awareness is to become quiet, listen to your soul and trust the information you receive. During meditation you can also move into your space of knowing. Ask questions and receive the information you desire.

Visualization

You too can experience the journey up the mountain to your higher consciousness and wisdom.

Spend several minutes focusing on your breathing and relaxing your muscles as you have done in previous chapters.

You may want to record a guided meditation for yourself, using some of the suggestions in this book. You can also add the following visualization to your focus and relaxation exercises:

> Walk, one step at a time, down 20 steps to a beautiful meditation garden. At the bottom you will find a particular path that seems to be drawing you to follow it. The path winds through rich, lush vegetation. You are surrounded by tall trees, sweeping vines, and beautiful plants. As you continue farther along the path, the expanse in front of you clears and you see a very large pyramid-shaped mountain. It looks about 1000 feet tall. You see that the path carefully weaves back and forth across the front of the mountain to make the ascent comfortable.

> You begin by heading off to the right and walking until the path switches back in the opposite direction, moving you toward the left side of the mountain and going all the way across the front of the mountain to the other side.

> Again the path switches toward the right. As you continue to move along, you stop for a moment and look back. You are now higher than the tops of the trees in the garden. The trail continues to move back and forth across the mountain until you find yourself at the top. The trail ends at a large wooden door in the side of the mountain.

> As you step forward, the door opens and you see the darkness inside. Slowly, carefully, you enter the cave and let the door close automatically behind you.

> You can sense a shift in the atmosphere inside the cave. It may be a bit cooler and very quiet. You feel safe and protected.

As your eyes adjust to the darkness, you notice some very dim indirect lights in the nooks and crannies of the cave.

You have come here seeking inspiration. As you wait in the quietness, looking around, some type of a messenger will appear. You may see it as a being or object, or you could hear it as a voice. It could even be something else that gives you the exact piece of information you desire.

As you trust that the information is here, you will receive your message or see your messenger. Allow the messenger to give you your message.

Do you understand what it means?

Tuck that message into your heart and its meaning will continue to deepen in your awareness.

Thank your messenger.

Note the acknowledgement back to you.

Turn and walk back to the cave door. It will open automatically. The sunlight from outside will be brilliant. It may take your eyes a moment to adjust.

As you now look out from your elevated position on the mountain, everything in your life can be viewed in better perspective. You look down at the garden and see the richness of the life within.

Nimbly you walk down the trail, feeling that with your new information, your life path has just taken on greater ease and purpose.

You return to walking through the garden. When you are ready, you can walk yourself up the 20 steps and bring yourself back, full of energy and brightness to your full waking state, treasuring your new information.

13

Become Your Soulution

Mastery of the spirit is the goal of becoming conscious. The physical world and the physical body are the teachers of our spirits.

What Do You Truly Desire?

What do you really want? Imagine having it already and how it will feel. Do you feel at ease with what you want? What is the feeling it gives you? Could you begin to enjoy it in your imagination? Why/Why not?

Is it stuff (material things) you want?

Basic necessities are important. What are they? In order to tap into them, why do you want these things, or in other words, how will they make you feel?

We give "stuff" a lot of power. Try focusing on the stuff you think you want and then ask how it will feel in your heart when you have it.

At this moment, most of you reading this book are probably just fine. Warm enough? Stomach full? In a safe shelter? You are able to direct your thoughts anywhere, to anything you want. Take those thoughts a step further and *feel the result of your thoughts.*

Imagine your perfect life. What does it feel like?

Is it possible to have a perfect life if you are wracked with disease? Or dis- ease?

Imagine and trust that wherever you are in your life right now is absolutely perfect for you.

Ask for What Is Rightly Yours

If you are asking for a new car because you want to look good or show off to your friends, filling this ego need may not be in your soul's best interests. If you are asking for a beautiful car that will ensure a comfortable ride and bring you much joy as well as a safe method of transportation, that is a different matter.

If you are asking for something that another person wants, it may not be your business. Jose, in the previous chapter, ran into a block when he thought he just wanted to make money. Once he tapped into reconnecting with his soul nature and intuition, his pathway opened up.

Energize Your Desires!

Think of your thoughts as vibrations that set into motion what you will create. Before your emotions can be meaningful to you—before they can give you the precise and perfect guidance they are offering—you must understand that **you are a being with two points of perspective that are continually relating to each other. Those two perspectives are: 1) your inner or higher self, and 2) your conditioned self (ego).**

Our higher self constantly calls us forward. We begin to understand our feelings of passion and eagerness when we listen to this expanded or higher self and allow it to draw us forward. Likewise, we understand the feeling of being unfulfilled or uneasy when we are not allowing our higher self to draw us forward. Uneasy feelings can be yellow or red flags for us that say **PAY ATTENTION!**

With awareness of your soul as the center, you can allow yourself to be the ever expanding being that allows you to feel joy. Contentment and joy tend to be like gratitude, constantly multiplying and delivering more of the same high-frequency energy.

Our emotions are indicators of the vibrational difference between *our point of desire and our current thoughts/beliefs and expectations.*

Point of Desire→**Emotions**←Expectations

Vibrational Difference

You must allow yourself to be the being that life (God or Source) is allowing you to become if you are to feel joy. **Unless you feel joy, you are not allowing yourself to be all you can be.** Living in the moment, allowing your internal Personal Guidance System to gently direct you, allows you to move with the flow or current of your life. Often in life we struggle. It may feel like everything we do is like paddling a boat upstream, against the current of life. Trying to move upstream is a lot of work.

Trust that your deepest desires and thoughts will unfold as you move forward with the flow of life as if you are moving with the current of life downstream.

Feelings of love, eagerness and joy are "downstream." Feelings of fear, anger and hate are "upstream."

Your perception of your current life experience may cause a realization that you do not have enough of something, e.g., money, time, knowledge, or stamina. Whenever you become aware of something that is lacking, you know more clearly what it is that you desire. **Immediately shift from what is lacking to what you desire.** If you are ill, for example, your desire for wellness will be clearly amplified. **Your focus needs to remain *only on what you do want.*** Your inner being flows with new desires. Every time you encounter something you do not want, it can help point you to what you really do want.

As your desires evolve all day, every day, the non-physical part of you also evolves because this is the part of you that will propel these new ideas into creation.

The more in touch you are with your inner self, the more effectively you can turn your undivided attention toward those new ideas. The positive rewards can be amazing. You will feel an eagerness for life, clarity of mind and vitality of the body.

In order to become fully in touch with one's inner self:

1. **Be present in the moment.**

2. **Keep your energy flowing freely.** Toss out excess baggage—unwanted thoughts, stories, feelings, attitudes—as soon as you become aware of them.

3. **Express your emotions.** Unexpressed emotions will be stored in the mind-body. If you allow yourself to experience and then release them, your body can stay in homeostasis or a state of balance.

Thoughts always precede manifestation. In the creation of all that exists, thought always comes first. Everything you see around you was once a thought or an idea, a vibrational concept that matured into what you call physical reality.

Our Will vs. The Highest Good

Soulutions are never about our ego or conscious mind. The ego operates according to our will and thus by nature is self-serving; **Soulutions are about intentions to do good for our soul self and for others.** It is the ego that often blocks us from achieving our full potential.

Help, My Entire Life Is Out of Control!

As a single mom in the days when I was raising five teenagers and a younger child, I can relate to several moments when I felt ready to shout to the world: **"Help, my entire life is out of control!**

When everything in your life feels out of control and you don't know where to start creating order out of all the chaos:

1. **Begin by taking a few calming breaths.**

2. **Choose one thing you can control.** It might be cleaning your closet or vehicle glove compartment, or even straightening a dresser drawer or emptying a wastebasket.

3. **Make sure that whatever you choose is visible and tangible.**

4. **Be aware of that first step** and what you have just accomplished. It is just a baby step; nevertheless, it is a step in the right direction. A small step, as tiny as it may seem, can give you a feeling of control.

5. From this sense of accomplishment, you can **choose another task that will help restore order**, and then another... and another.

6. Now, quiet yourself. **Move your awareness deep inside of you. Identify something important that has significant meaning for you that you can control.**

At times, the only thing within your control is your attitude. These are times when it can help to remember that we are humans on a soul journey.

Years ago, I put my life on hold and spent a month with my dying mother. After her funeral, I arrived home late on a cold winter night looking forward to getting life and home back to normal.

The next morning before dawn, I arose and said to myself as I headed downstairs, "Today my life will get back to normal."

As I stepped off the bottom stair into the living room, I sank into several inches of icy water. For several moments I just stood there, my feet submerged in the water and letting it register in my mind that the downstairs of my home had somehow flooded.

As I stood in the cold water I said out loud to myself, "This is Normal."

At that time, as a single mother with a brood, the only thing I could expect was the unexpected. I must have stood at the bottom of the stairs for 15 minutes trying to figure out how to transform this flood into something good.

It was three weeks before Christmas. What would be the best way to create something good out of this that would work well for me? That was the intention I set for myself, so I continued to search every corner of my mind until I came up with a plan.

First, I turned on the lights and surveyed the damage. As I gazed around the room, I decided it may not have been such a good idea to turn on the lights while standing in the water. Oh well. The carpet was old and would have to be replaced. We would have to cut into the walls in order to drain the water from the room.

My heart lifted. What a great opportunity for a minor remodeling! I'd get rid of everything that needed to be replaced or that I'd never liked anyway!

I called the insurance company, woke up the children and then the cleanup began. In the midst of grief over losing my beloved mother and the children's grandmother, we launched into a month of working together as a family. The family room where we normally would have spent Christmas was now stored with furniture. On Christmas Eve, we rented a storage unit, moved the furniture and purchased a small tree.

Actually, it turned out to be one of our best Christmases. I had no decorating to do, thus, no Christmas clean-up. The kitchen and dining room were still intact, so we had a normal holiday dinner. Using the insurance money, friends helped us knock out a wall to create more open space. We didn't replace the carpet. Instead, I stained the concrete.

Two months later I had a house I loved more than before. I realized that all of this came about because I had made the conscious choice while standing in that icy water in the dark of dawn, that I was going to figure out a way to turn the flood into something that worked well for me. We were all safe, we grew closer through work, and the children felt needed, expecting less from me.

I was back to my normal life all right—with a houseful of teenagers and a little one and all the usual chaos! "Normal" was never going to mean outwardly calm and predictable. Rather, **normal for me meant keeping my calm and peace internally.** I actually got myself a license plate that reminds me to *LIVCALM.*

What Do We Do with the Truckload of Mistakes?

How can we use past mistakes to best serve us as we move forward on our soul path?

By now we may know what we want and we are aware also of what doesn't work. Clearly we can ignore those unrewarding thought patterns and give ourselves permission to replace those thoughts with ones we really want.

A phrase I find helpful to release old patterns and thoughts sounds something like this: "Up until this time I_____; now I _____."

For example:

"Up until this time I have struggled in relationships; now I tend to draw to me people who are a good match."

"Up until this time I habitually ate too many sweets; now I choose those foods I love that serve my body well."

"Up until this time I have struggled with taking tests; now I recognize after I study that I know much more information than I will be asked to recall."

"Up until this time I thought all my problems were caused by my dysfunctional parents; now I recognize I can choose my thoughts."

Barring any major personality disorders, by adjusting our thinking and sometimes with outside help, we will be able to get our life under control.

At times, our attention is placed on controlling things on the outside of us. **Ultimately we want to live by allowing the deepest part of us to control our thoughts from the inside**. We then can feel the peace and contentment of knowing we're on our soul path.

Principle Based Living – Align Your Thoughts & Feelings with Action

I work with many people whose childhoods were extremely challenging. Some were either molested, abused or abandoned. Their subconscious and conscious minds—their "operating systems"—were filled with false messages of who they are and what they can expect of life.

With a faulty operating system, they continued to experience what their downloaded programs—their beliefs, feelings, attitudes, outlooks, behaviors and actions—delivered to them. Predictably they were often consumed with worry, fear and doubt.

Even in a great home with loving parents, children pick up false messages, or they perceive things differently from the way they were intended. Everyone has emotional work to do to benefit them and help them grow.

If we operate only according to our installed programs, we will continue to see the world through lenses of the past. It is inevitable that some people will perpetuate the cycle of dysfunction they grew up with. Fortunately, many ask for help or decide to move forward.

Align Your Thoughts & Feelings with Action

Have a "brain conference" to align your goals and dreams with your behavior. Begin by asking: How do I love and accept myself in the present? Loving oneself and knowing you are loved is essential to being able to love others. Can you look into a mirror and say "I love you" to your image? That is a good place to begin.

According to Charles Darwin as paraphrased from *The Origin of Species*, it is not the most intellectual of the species that survives; it is not the strongest that survives. The species that survives is the one that is best able to adapt and adjust to the changing environment in which it finds itself.[80]

As humans we find ourselves in continually shifting environments. As we courageously let our hearts lead us, we will not only survive, but thrive.

There is a difference between being interested in changing and being committed to change and growth. If we are interested, we will do it if it is convenient. Otherwise, we will make excuses not to do it.

Commitment to change requires investment in ourselves.

What is the level at which you want to commit to your personal growth? What are you willing to do to upgrade your skills? This awareness and commitment to soul growth is a lifelong journey.

Gratitude Is Everywhere

Recently, I attended an event called "The Brittania" in Beverly Hills. It was similar to the Academy Awards but for British actors and producers in Hollywood.

Growing up in Burbank, California, movies and movie stars were part of the everyday landscape and were not very exciting to me. I felt like the entire movie scene was a lot of hype and ego.

How surprised I was when the awards were handed out and the recipients were gracious, kind and grateful. They all heaped the praise for their award on those who supported them, downplaying their own part. One of the recipients expressed much gratitude to her directors, giving them credit for her success. Another recipient who hosts a late night CBS show showered high praise on his employers and his lovely wife for her support.

---

[80] http://quoteinvestigator.com/2014/05/04/adapt/

Most impressive was a recipient who was honored for his international work with UNICEF. He travels the world as a spokesperson for UNICEF, visiting some of the most impoverished high-risk places, spreading love and connection. During the Ebola outbreak, he was on the scene. He was also shown in a video helping Syrian refugees and others in need. This man's passion is connecting with the children. He heaped praises on the parents of the suffering for holding their families together. His platform was clearly gratitude for being in a position to work with some of the most challenged in this world. One could feel his awareness of how honored and privileged he was that his status as a celebrity enabled him to render so much service to others, even at the risk of his own health and safety.

Those stars shone that night. I came away with a shifted perspective of the Hollywood I remembered as a young girl.

Divine Energy

Divine energy is inherent in our biological systems. Every thought that crosses our mind, every belief we nurture, every memory we cling to, translates into positive or negative commands for our bodies and spirits.

It can be magnificent to see ourselves through this lens, but it is also intimidating because no part of our lives or thoughts is powerless or even private. We are biological creations of Divine design. Once this truth becomes part of our conscious mind, we can never live an ordinary life again.

Our Energy Is Our Spirit

Carolyn Myss says she uses the word "energy" because it is more neutral and evokes no religious associations or deeply held fears about one's relationship with God. She says it is easier to tell someone, "Your energy is depleted," than to tell them, "Your Spirit is toxic."

Seeing a problem through a spiritual lens accelerates healing because it adds a dimension of purpose and meaning to a person's crisis.

Our spiritual task in this life is to learn to balance the energies of body and soul, thought and action, physical and mental power. Our bodies contain an imminent blueprint for healing.

How would your life be enhanced if you infused heightened consciousness of the Sacred within into your day to day life?

As we become more conscious and recognize the impact of our thoughts and attitudes—our internal life upon our physical bodies and external life—we no longer need to conceive of an external parent God that "creates for us," on whom we are fully dependent. As spiritual adults, we accept responsibility for co-creating our lives and our health. **Co-creation is, in fact, the essence of our spiritual adulthood.** It is the exercise of choice and the acceptance of responsibility for those choices.

Our Divine contract is management and acceptance of responsibility for our power choices.

Intuition: What Part Does It Play?

The "inner knowing" and "soul awareness" we have been talking about throughout this book may be labeled by some as intuition. Intuition is simply the ability to know and understand without conscious reasoning or external validation. Intuition is that inner voice that taps into the guidance from your subconscious mind and your soul.

As you practice the exercises in this book and start to meditate if you haven't already added this self-time to your daily life, you will develop your intuition and step into greater alignment with your soul path.

Some people think intuition is genetic. I disagree with that. Although sensitive individuals may have easier access to their inner PGS, **everyone can develop and increase their ability to listen and follow guidance.** Meditation is an effective way to develop and expand our intuitive abilities.

I have found that sometimes it is the most highly sensitive people who medicate with alcohol, cigarettes or drugs because they are bombarded with information and feelings. When those individuals release their unhealthy coping mechanism, they can greatly heighten their awareness.

Rosa

"Rosa," a successful businesswoman, presented in my office with an alcohol addiction. She recognized that she had been having periodic "blackouts." She was frightened because she had absolutely no memory of her actions during those times. We worked through some of the underlying family issues the alcohol had been masking.

As Rosa regained her gift of sensitivity she started to study and soon became proficient with energetic healing techniques. She began to work with people who were physically ill and transformed herself from a medicated addict to a gifted, sensitive healer.

If you feel like you are not receiving intuitive guidance, maybe it is blocked by old beliefs or programs. In that case, start peeling back the layers of insulation, dumping old baggage, and releasing those unwanted stories about yourself.

You may begin by setting your intention to live more soul-centered. I suggest writing an affirmation to yourself, declaring such ideas as:

- I am extremely intuitive.
- I have every answer I need inside of me already.
- I can clearly hear the voice of my intuition.

- My intuition guides my life every step of the way.
- I am open to the many ways intuitive guidance comes to me.

Intuition or soul guidance may appear as an inner voice, a slight feeling in your heart or gut, an image or symbol in your mind, or a deep knowingness.

As you open your heart you increase your intuition. Raising the vibration of your heart through the emotion of love is a powerful tool to access your highest self and also connect with other souls. The work and exercises we have discussed thus far in this book are a guide to further opening our awareness.

HeartMath Institute's studies show that the heart radiates with an electrical current 40-60 times more powerful than the brain's electrical emissions, and the heart is magnetically over 1000 times more powerful than the brain. Their research shows that when we feel genuine (heartful) intentions of desire, electromagnetic patterns are formed in our hearts. Their hypothesis is that conscious creation influences templates in our molecular structure that amplify the new pattern and then harmonize and mirror the earth's magnetic frequencies, allowing us to manifest our intended desire.

Breathing with intention and staying present in the moment can assist us in remaining open to inspiration.

"Live Pure"

How can you know if the information you receive is from a source of light? What does it feel like?

To ensure that all of your intuitive information is from a source of light and truth, "live pure." Pure is defined as: 1) not mixed or adulterated with any other substance or material, 2) without any extraneous unnecessary elements, 3) free of contamination.

Pure is simple. A pure heart is true to its nature. **We might say that living pure is living in truth as who you really are, uncontaminated by false messages.**

I Will Be Judged

Six years ago I made my own big life-changing decision. I absolutely knew it was the right thing to do, but I also knew some would judge me harshly. After a year of pondering, meditation and prayer, the morning I followed through on my inspiration, I awoke to a loud beautiful chorus of rejoicing angelic music in my mind and deep peace, confirming to me that I had done the right thing. More surprising than that was that EVERY morning for one year I was awakened the very same way. I felt the angels rejoicing with me for my courage. I had a full year of waking up to joyous music and confirmation of my choice.

Have you ever mysteriously had a melody playing in your head? It may be one you haven't heard in a long time. Listen for words behind that message. You may receive an answer, and that melody could be the sound of your intuition.

Transformation Doesn't Mean Starting Over

Our thoughts are continually creating the body chemistry that produces our feelings. **Control your thoughts and give your emotions and your body the proper direction by carefully choosing the most effective words to convey your message.** Watch how the art of moving your life in the direction you want becomes like turning a large ocean cruise ship. The movement will be slow and steady.

I tell my clients that changing their lives is more like a shift than a change. It feels so much easier to shift a gear in a car than to stop, turn around and reverse your direction. Even the thought of starting over can be overwhelming. Shifting, on the other hand, simply requires moving a few degrees in a new direction. **Only a few degrees of change in our thinking can lead us to an entirely different place in our awareness. That shift could**

be in the words we choose to tell ourselves about what it is that we really want.

Some will ask, "What about really big changes? Suppose we perceive we've messed up our entire life?"

First, realize you don't want to keep doing things the way you did them before. The past is simply that, the past. Let it just become history. We study history to make better current decisions. We want to improve the quality of our life by doing things with more clarity and direction. **The shift will happen naturally if we stay focused, moment by moment, on our new way of thinking and communicating with ourselves.**

Use Spiritual Protection

Some information or ideas are not "of light" or truth. Always ask for protection from impure or untruthful information and then expect that protection.

If you are not "living pure" and are involving yourself in harmful or destructive practices, then you may open yourself up to energies that are not Divine.

Asking for spiritual protection can be as simple as requesting to be surrounded by Divine light, Divine energy, beings of truth, angels, or even white light. As you ask for this protection, visualize it in your mind. It will be there.

How Does the Truth Resonate with You?

How do you know what is rightly yours to ask for? How do you know what is not rightly yours? When you ask questions of your deeper self, what is behind your intention?

What is the feeling that comes to you when you receive or retrieve information? It should be loving and calm. It should also feel accurate. If you are getting unsettling answers, ask more questions until you gain understanding and peace. Ask for the courage to follow through on your truth.

Live in truth and honesty in all aspects of your life. Ask for guidance.

Synchronicity

Dr. Schwartz writes about synchronicity or "meaningful coincidence." He says often it is Spirit trying to get our attention. The more you become aware of synchronicities and the more you ask and listen for them, the more frequently they seem to manifest.

Examples of synchronicity are:

1. You're thinking about someone and they call, email or physically show up. If they've already passed over, they may appear in a dream.

2. You visit a certain place and it is so familiar, you know you've visited it before. In fact, it's so familiar you can give the directions to certain houses, stores, parks, etc. You cannot explain how you know this information.

3. You meet a person whom you've never met before, but you are keenly aware that you already know them.

4. Someone talks about a certain event in history and although you know nothing about this event, it stirs a memory deep inside. Suddenly these memories start to surface and the event is as vivid to you as if you had personally witnessed it.

5. You see a series of coincidences that point you to a solution you have been seeking.

We have so many unanswered questions regarding synchronicities. Why do they happen? Could they be a message from beyond? Why would Spirit choose to communicate in this way? What can we learn from these experiences?

Here are a few of my thoughts in response to those questions. I think synchronicity occurs sometimes just to perk up our awareness, almost as an energy booster that causes us to pay attention and sharpen our sensitivities.

Synchronicities may happen to point us in the direction of healing for our body, mind or soul. It could be Spirit's way of getting us to pay attention.

When synchronicities occur for you regularly, they can validate that you are on your authentic path. They can be a beautiful confirmation that you are paying attention.

The Butterfly that Saved the House

"Judy" came to me seeking soul clarity. She had a strong history of trusting that she would get the information she needed when she needed it. She told me recently that she had bought a house. She had been unsure as to whether this house was the best one for her in order to restart her new life.

Judy said that whenever she needs confirmation of a decision she seems to get answers from butterflies. Sometimes it is through seeking a picture of a butterfly or hearing someone talk about them. She said her father loved butterflies and she felt like some of those answers were a sign from him. Her mind was primed to see butterflies.

Judy was standing alone in the foyer of her prospective new home. She was quieting her mind and lifting her heart, asking whether "this was the house to buy." As she looked upward into the skylight of the home, what did she see but a butterfly flying high above her head! She bought the home.

Judy recognized the synchronicity of the butterfly in a home that had been locked up for days. She wondered how the butterfly got there but heeded the message. It turned out to be a perfect purchase for her.

Pay attention to the little things. Sometimes they are the big things, as big as a house!

Time & Place

> *The right place at the right time, the wrong place at the right time, right place at the wrong time, it all works for you.*

How many times have you heard someone say, "I was at the right place at the right time"?

Or: "I was at the wrong place at the wrong time"?

We love those great lucky breaks that come our way: the steal of a deal in a car or a home, the lucky half-court basketball shot in the last three seconds of the game. We can clearly see this is a good thing for us.

Sometimes it is much more difficult to accept what we perceive as those wrong place/ wrong time circumstances. We do not like to be turned down for the job we want, or lose our house in a flood, or miss an opportunity.

Yet when we look at all circumstances in our life from our soul perspective, those losses and disappointments can become our greatest teachers. They can help shape our character, strengthen our faith, and evolve our souls. Unwanted circumstances can allow us to delve deep within ourselves and discover creativity. They can motivate us to shift priorities and focus on gratitude for all the things we do have. Sometimes when all is lost, we discover our real character, our priorities, and our purpose.

Most people have loved ones or friends who, when confronted with a life-threatening illness, walk into it with grace, ease and peace. It is during times of the greatest challenges that we are forced to turn inward to the deepest parts of ourselves.

Living (En)Joy

We use the word "enjoy" to describe how we feel about many activities or experiences that are pleasurable. The word "enjoy" is a regular part of our day to day conversation. We enjoy a good meal, we enjoy a vacation, we enjoy a good book.

Although pleasurable activities are nice, **living from our heart in the emotion of joy is much deeper and more satisfying.** I believe we can move the meaning of the word "enjoy" deeper to our heart and soul level. For example, I'm not sure our soul actually enjoys eating a good meal—or whether we like the taste of the food and the fact that it brings us pleasure (a fleeting sensation). If, however, we are en-joying our meal in the company of people we love, then the memory of that event as well as the delicious taste of the food will enhance the meaning of enjoyment.

Living (en)joy, centered in our heart, affects every cell in our body at a physiological and emotional level. As you regularly check into your heart, bring the feeling of the present moment to that higher state of "en-joyment." Immerse yourself "en-joy." Not only will the moments be brighter, but your health will also improve and you will intensify the feeling of gratitude—a feeling that is so self-replenishing.

Make Time for Heart-Filled Fun

Do you remember how much fun you used to have as a child—maybe running into the wind and holding out your hands to feel it sifting through your fingers? Or catching snowflakes on your tongue? Building snowmen and snow forts? Searching in the grass for four-leaf clovers, or sloshing through the crisp ground layer of colorful leaves and listening to them crackle beneath your feet? Picking wildflowers and pressing them between the pages of a book?

Or maybe the fun you had at the beach, splashing into the waves and trying to catch the uppermost crests before they came crashing down over you... collecting beautiful colored stones, some with fossil imprints from sea animals... creating sand sculptures... paddling a canoe or kayak, or possibly tipping the canoe and forming an air pocket underneath where you could talk...

Or maybe inside at home, kneading bread dough and making cut-out cookies or tea sandwiches...

If you grew up in a climate where it snows in the winter, maybe you loved to go tobogganing or ice skating with your friends. Oh, the wild feeling of speeding down a snowy slope, holding onto each other, yelling and screaming excitedly... or, gliding smoothly on the ice, hearing the scraping sound of your skate blades as you spin around and around... Then later, warming up in front of a blazing fire and sipping a steaming cup of hot chocolate...

In the spring, maybe you enjoyed biking over to one of your friends' houses where everyone gathered "just to gather." What did you talk about? You can't remember? That's okay because it didn't really matter. It was just fun to be together.

Maybe you loved to camp out with your family, or go on vacation to the mountains or to another beautiful spot in nature. Maybe sometimes when it was raining outside, you loved to sit on the porch of a mountain cabin with your parents or sisters and brothers doing a jigsaw puzzle that had hundreds of tiny pieces. Or maybe you played board games together.

The more you delve into your memories, the more fun times you will discover. Maybe your favorite times were just being alone, reading a book or doing crossword puzzles. Capture those feelings of joy and contentment and bring them forward into your life today.

Die Consciously by Living Consciously

Dying consciously is one of the many blessings of living consciously. How can we work with our minds to refine our perceptions and become skilled at penetrating illusion? It takes discipline and patience to master anything, so it is important to remember at all times that we are a work in progress. **We cannot always measure our success. It is not linear.**

Often we have deep doubts or dark nights of the soul. Other times, we have great spiritual leaps forward. Symptoms of the dark night are feelings of lack or loss, such as absence of meaning or worth, or loss of self-identity.

A person does not look to others to blame but looks inside themselves, knowing their crisis is sourced from within.

When the Student Is Ready, the Teacher Appears

For years, I always planned to formally study medical intuitive work. I figured I would have to wait until my youngest child was out of the house because I thought all of the programs were thousands of miles away. Yet despite feeling like I would have to wait, my longing was so great, the very next year I decided I would begin the process of preparing for formal study.

New Year's Eve is usually a time when I like to stay home and enjoy a quiet celebration. That year, however, I decided to attend a party. Imagine my surprise when I met my medical intuitive teacher at the party! I began formal study five days later and continued training with her for two-and-a-half years.

In retrospect, I found it interesting to discover that this particular window of time was the only one I could have worked into my busy schedule with two children still at home.

I did not formally call out for a teacher, but apparently the longing of my heart was so great, it was definitely in my soul's best interest to bring that teacher to me at that time.

It is important to trust and know that teachers appear when we truly want them and are ready for them. Teachers allow us to grow and become teachers to others. Curiously, **sometimes our adversaries become our best teachers.** Those who provoke strong emotion within us can cause us to look deep inside of our psyche and grow.

Prayers Are Always Heard & Answered

It had been a stressful 10 days. I had a sweet client, "Marcia," a woman in her seventies, who had been having issues with her colitis for several years. Ultimately I ended up taking her to the ER and serving as her medical advocate. My responsibilities to this woman, in addition to my full-time work and family life, left me feeling exhausted.

Through a series of events at the hospital, I knew it was imperative that I pay close attention to this woman's medical care. Because of my own nursing experience in having witnessed too many errors—most recently the death of my beloved mentor—I felt a tremendous responsibility.

Each night when I finally collapsed into bed, I heard angelic music. This was not unusual for me. Previously in the most stressful periods of my life, especially in the past few years, I've heard the soothing musical messages of faith, hope, love, and reassurance.

The day Marcia was released from the hospital I got the message in words: "Your prayers are answered." The message was crystal clear. I knew it was more than merely about my client's safety or healing.

I wondered, "What prayer?" I had so many. I wanted answers and I wanted them now!

Shortly after receiving that message, my 29-year-old daughter, single mother of two children, for whom I say a lot of prayers, returned to school and in a frenzy of activity with much inquisitiveness and determination, began to study everything she could about her own soul and her path. I witnessed my beautiful, intelligent, gifted, compassionate daughter beginning to use her gifts to help others and enrolling in a course to study Counseling Psychology.

A son who was injured on the job three years prior and only a month after getting married, began studying in a program to redirect his career.

My college athlete son who had moved away to be on the basketball team broke three bones in his foot and had to completely restructure his life to one that was service-oriented versus the former ego-oriented sports world.

I believe because I served where I was needed, I received extra assistance for my own family needs.

It is so important for our souls to make life and service choices out of love and compassion, not to earn brownie points or blessings. Our choices are founded on the simple fact that **we want to make a difference.** We surely can.

My father was a kind, compassionate, patient and faithful man. He always taught me, "You cannot out-give God." I have found this to be true. **The more we seek to give to others, the more we receive in return.**

Walk Your Talk - Your Soul Knows the Difference

Is your talk "your walk," or is it the voice you've learned from someone else?

Why is it so common to not walk our talk? Simply because we're human. We don't like to rock the boat, risk rejection, or perceive failure. Sometimes following our soul path means speaking up for a cause that can result in being subjected to backlash. Other times, authentic, heartfelt expression might lead to leaving a relationship or risking rejection.

Let your soul help you walk your talk. Delve into that deep knowing aspect of your nature and listen to what it is telling you. **Your heart and soul are always your best guides.** Your soul will lead you to deeper places of inspiration and guidance.

We *Can* Take It with Us

How many times have you heard "you can't take it with you" in reference to placing too much value on material goods? There are some things of great value that I believe we do take with us. That is where our investments pay off big time.

We can always take soul learning with us. The soul's impressions or ideas once ignited are unstoppable, regardless of the odds. This is why we want to shift our minds into higher gear and discover *soulutions*, not just temporary "solutions." Everything we experience, physical and emotional, affects our spiritual self.

The Ease of Transition

Sometimes we hear people talk about their "next life." What is our next life if not a continuation of our present one?

The journey of our soul is continuous. **Our next life is really just an ongoing continuation of our current soul life.**

As I was discussing soul issues with a friend of mine, Terry Muech, an executive life coach, he mentioned the term "Optimal Incarnation" to me. I latched onto it immediately, as it is a complete description of what all of us can seek and experience in this precious lifetime.

Terry and I were talking just as I put the finishing touches on this book. I thought if there was a take-away from the science and stories of what you have read, it is just that: **to increase awareness and seek to live an optimal incarnation.**

I believe everything we have discussed in this book points us to recognize and embrace that theme. **As eternal spirits in human bodies, awareness and knowledge of science of soul and heart can help us keep our own soul as a focal point in every choice and decision we make.**

Over the centuries, science has expanded and enhanced our way of life. Soul awareness has been around since the beginning of time, and now soul science is catching up and helping us make the leap of faith less of a leap and more of a choice.

Life lessons focused on learning and expressing love and compassion are treasures we take with us on our ongoing journey. **Overcoming our natural human struggles with anger, jealousy and war are also precious soul lessons we take with us.**

We are witnessing a gigantic shift in consciousness everywhere on our planet. I believe as we embrace who we really are and focus our energy on evolving that most enduring part of us, we will be living our unique role of expanding peace and love.

As we recognize the "soup of consciousness" of which all of us are a part, we will understand that **when we raise our own vibration we also raise the vibration universally.**

My wish for each of you is to embrace life—all of it: the joy, the struggles, the triumphs and defeats, especially the love and compassion, so that during your last days in your body you can say with peace and in gratitude, "My life was an optimal incarnation!"

Questions & Reflections

1. As an adult, what do you do for fun or just because you enjoy it? Make a list of all the fun things you used to do. Do you prefer being inside or outside, on land or water, moving or still, alone or with someone else—or maybe with several others? If you're stuck and can't really make your list, consider what used to make you feel alive. It might be something you've forgotten about. Now reclaim that memory again; it will suddenly start to light up your heart with feeling.

2. How do you express your creativity?

3. What brings you your greatest joy? Explain.

4. How do your creativity and joy play into your soul path adventure?

5. If you could improve one aspect of your life at this time, what would it be? What is preventing you from making that change now?

Visualization

Begin with attention to your breath. As before, spend several minutes calming your body and quieting your mind.

Become aware of your body. Allow the muscles of your feet and toes to soften and relax. Let that comfortable feeling wash up your legs and right over your kneecaps like an ocean wave washing up on the shore. Then allow a bigger wave to wash from your knees all the way up to your hips and pelvis.

As the relaxation that started in your feet moves up into your hips and pelvis, imagine the bones of your pelvis actually softening as the muscles relax; it is such a comfortable feeling.

Gently ease the comfort up through your abdomen to your diaphragm, letting the feeling of relaxation lift and rise, like warm air rising up in a room. Move the easy feeling of release all the way up to your shoulders.

Allow your shoulders to feel like a gentle warm waterfall is washing over them and flowing freely down to your elbows. From your elbows the comfort continues down to your arms, through your hands and fingers... running through your fingers as easily as water running downhill.

Smooth the muscles around your throat. Imagine the actual diameter inside your throat is gently opening, the way an aperture of a camera opens to allow more light. As you open your throat you are opening the passageway between your heart and your brain. The messages from your heart can easily travel upward to your awareness through your open and relaxed throat.

Imagine you can look inward toward your heart and see a beautiful golden light. That is the light of Love flowing from your heart. Feel its warmth... notice the streaming rays of light moving outward from your heart, allowing you to see everything around you with more clarity and brilliance.

Your inner eyes and inner ears are open. Easily you can see and hear the true meaning and messages of all situations and circumstances you encounter. You feel calm. You are at one with the world.

Imagine that the loving light emanating from your heart is illuminating all that is around you. You begin to see everything more clearly in a loving, kind and compassionate way.

Tune in often to your heart light.

Infinite Possibilities

Just as our souls are infinite, so is the universe. Imagine the joy in knowing we have infinite lifetimes to experience joy and abundance, and infinite possibilities to create new ways of seeing and being!

When we immerse ourselves in the essence of what this truly means, every aspect of our current life looks and feels different. How fortunate we are to be living in an information age that delivers copious amounts of data scientifically validating the existence of our soul and its connection with all other souls in the cosmos. If we can flip a switch to turn on an electric light fixture, then we can also flip a switch to open our hearts and flood our entire being with enlightenment.

Each of us has every bit of information we need inside of us. As we learn to access it, we begin to blossom into fullness like a beautiful blooming rose in springtime.

Science of the soul can open your mind to limitless possibilities.

I see each day as a gift to experience this life in our body. We can make choices, expand our capacity to love, and shed the parts of human nature that we want to release.

Living in this awareness makes each moment significant. In spite of the turbulence of the world, knowing my soul's purpose brings a deep sense of calm to every part of my being. It is the same for those with whom I work as well as for the many other practitioners who use similar techniques for helping their clients achieve clear communication with their soul or superconsciousness.

Begin today to create your own Optimal Incarnation. Use the tools we have discussed, take time throughout the day to breathe and meditate. Daily life will have more meaning as a new calmness washes through your being. Your heart will expand as your soul leads the way.

Listen to the messages you receive, and pay attention to the road signs along your soul path. Do take time to smell the roses and look for four-leaf clovers or falling stars. Or... just let them appear on their own. They will, if you let them. Synchronicity has a way of showering us with delicious surprises.

Laugh often and let others laugh with you. We are so full of life and love, there's no room for sorrow or loss unless we can transform it into an act of gratitude for the deliverance of yet another learning experience.

Our lives are an amazing adventure. Let yours be an Optimal Incarnation!

Mary Kay Stenger

Dr. Mary Kay Stenger has over 35 years of experience in integrative medicine as a certified Medical and Clinical Hypnotherapist, Nutritionist, and Spiritual and Intuitive Healer. She is also a Registered Nurse with experience in Intensive Care and Emergency Medicine, and has taught as a Nurse Educator. Dr. Stenger has a Doctorate in Psychology with associated advanced degrees in Health and Preventive Medicine and Health Education Psychology. In 1981, Dr. Stenger was a nurse on the Reagan/Brady medical team in the wake of an assassination attempt on President Ronald Reagan.

In her private practice, Dr. Stenger uses powerful intuitive and energetic healing techniques to create a safe environment that fosters a renewed healthy balance of body, mind and spirit. She has developed many highly effective programs in a wide variety of health and motivational areas. Her techniques focus on creating life balance in order to uncouple destructive issues rooted in the subconscious mind.

For corporate clients, athletes and public organizations, her techniques effectively improve efficiency and performance while enhancing employee health, morale and work relations and lowering associated business operational costs. As clients and organizations move beyond their limits, they excel in all areas of life.

Dr. Stenger is the founder of Tender Hearts, a support group which serves multiple birth families that have lost a child. Her research in the field of grief was published internationally and she was honored as keynote speaker in Amsterdam at the International Association of Twin and Multiple Birth Studies. She has also served as Medical Adviser to the former Hypnosis

Training Institute of Central California. Dr. Stenger is the mother of six children, including three surviving quadruplets.

Dr. Stenger's website for more information and for announcements of upcoming books is www.soulpathsolutions.com.

Bibliography

Alexander, Eben, M.D. *Proof of Heaven: A Neurosurgeon's Journey into the Afterlife.* New York: Simon & Schuster, 2012.

Beck, Martha. *Finding Your Way in a Wild New World: Reclaim Your True Nature to Create the Life You Want.* Atria books, 2013.

Burney, Robert. *Codependence The Dance of Wounded Souls: A Cosmic Perspective of Codependence and the Human Condition.* Joy to You & Me Enterprises, 2012.

Cheek, David, M.D. and Rossi, Ernest. *Mind-Body Therapy: Methods of Ideodynamic Healing in Hypnosis.* New York: W. W. Norton & Company, 1994.

Childre, Doc Lew and Wilson, Bruce, M.D. *The HeartMath Approach to Managing Hypertension: The Proven, Natural Way to Lower Your Blood Pressure.* New Harbinger Publications, 2007.

Childre, Doc Lew and Martin, Howard. *The HeartMath Solution: The Institute of HeartMath's Revolutionary Program for Engaging the Power of the Heart's Intelligence.* Harper One, 2000.

Childre, Doc Lew and Rozman, Deborah. *Transforming Anxiety: The HeartMath Solution for Overcoming Fear and Worry and Creating Serenity.* New Harbinger Publications, 2006.

Chopra, Deepak, M.D., *Quantum Healing:Exploring the Frontiers of Mind/Body Medicine.* New York: Bantam Revised & Updated, 2015.

Cousins, Norman. *Anatomy of an Illness: As Perceived by the Patient.* New York: W. W. Norton & Company, 2005.

Dali Lama. *The Heart of Meditation: Discovering Innermost Awareness.* Shambhala, 2016.

Descartes, Rene (author) and Donald A. Cress (translator). *Discourse on Method and Meditations on First Philosophy, 4th Ed. 4th Edition.* Hackett Publishing Company, 1999.

Emoto, Masuro. *The Hidden Messages of Water.* Atria Books, 2005.

Gerber, Richard, M.D., *Vibrational Medicine: The #1 Handbook of Subtle-Energy Therapies (3rd Edition).* Bear & Co., 2001.

Hay, Louise. *You Can Heal Your Life.* Carlsbad CA: Hay House, 1984.

Hay, Louise. *Life Loves You: 7 Spiritual Practices to Heal Your Life.* Carlsbad CA: Hay House, 2015.

Katie, Byron and Mitchell, Stephen. *Loving What Is: Four Questions That Can Change Your Life.* Three Rivers Press, 2003.

Kubler-Ross, Elisabeth, *On Death and Dying: What the Dying Have to Teach Doctors, Nurses, Clergy and Their Own Families. New York: Scribner's reprint, 2014.*

Lipton, Bruce. *The Biology of Belief: Unleashing the Power of Consciousness, Matter, & Miracles (Revised).* Carlsbad CA: Hay House. 2007.

Moody, Raymond, M.D. *Life After Life: The Bestselling Original Investigation That Revealed "Near-Death Experiences".* New York: Harper One (Special Edition), 2015.

Moorjani, Anita. Dying *To Be Me: My Journey from Cancer, to Near Death, to True Healing.* Carlsbad CA: Hay House, 2014.

Myss, Carolyn, Ph.D. *Anatomy of the Spirit: The Seven Stages of Power and Healing.* Harmony, 1996.

Newton, Michael, Ph.D. *Journey of Souls: Case Studies of Life Between Lives.* Llewellyn Publications, 1994.

Newton, Michael, Ph.D. *Destiny of Souls: New Case Studies of Life Between Lives.* Llewellyn Publications, 2000.

Pert, Candace, Ph.D. *Molecules of Emotion: The Science Behind Mind-Body Medicine.* New York: Simon & Schuster, 1999.

Pert, Candace, Ph.D. *Molecules of Emotion: Why You Feel the Way You Do.* New York: Simon & Schuster, 1998.

Schwartz, Gary, Ph.D. (Author), Chopra, Deepak, M.D. (Foreword), Simon, William L. (Contributor). *The Afterlife Experiments: Breakthrough Scientific Evidence of Life After Death.* Atria Books, 2003.

Schwartz, Gary, Ph.D. *The Sacred Promise: How Science Is Discovering Spirit's Collaboration with Us in Our Daily Lives.* Atria Books, 2011.

Van Praagh, James. *Healing Grief: Reclaiming Life After Any Loss.* New York: New American Library, Penguin/Putnam, 2001.

Weiss, Brian L., M.D. *Many Lives, Many Masters: The True Story of a Prominent Psychiatrist, His Young Patient, and the Past-Life Therapy That Changed Both Their Lives.* Fireside, 1988.

Weiss, Brian L., M.D. *Only Love Is Real: A Story of Soulmates Reunited.* Grand Central Publishing, 1997.

Wonson, Beth. *Let Go of the Rock! A New Look at the Dynamics of Self-Management.* Mesa AZ: Dandelion Books, 2015.

Wonson, Beth, Stenger Mary Kay, Ph.D., *An Everyday Guide to Joy & Abundance.* Mesa, AZ: Dandelion Books, 2016.

Zukof, Gary. *Seat of the Soul.* New York: Simon & Schuster, 2007.

CPSIA information can be obtained
at www.ICGtesting.com
Printed in the USA
FSOW01n1853231116
27650FS